The Clinical Pharmacology
of Biotechnology Products

THE CLINICAL PHARMACOLOGY OF BIOTECHNOLOGY PRODUCTS

Proceedings of the Esteve Foundation Symposium IV,
Son Vida, Mallorca, 7–10 October 1990

Editor:

M.M. Reidenberg
Department of Pharmacology and Medicine
Division of Clinical Pharmacology
The New York Hospital – Cornell Medical Center
New York
USA

 1991

EXCERPTA MEDICA, Amsterdam – New York – Oxford

International Congress Series No. 944
ISBN 0 444 81376 4

Published by:
Elsevier Science Publishers B.V.
(Biomedical Division)
P.O. Box 211
1000 AE Amsterdam
The Netherlands

Sole distributors for the USA and Canada:
Elsevier Science Publishing Company Inc.
655 Avenue of the Americas
New York, NY 10010
USA

This book is printed on acid-free paper.

Library of Congress Cataloging in Publication Data:

Esteve Foundation. Symposium (4th : 1990 : Son Vida, Spain)
    The clinical pharmacology of biotechnology products : proceedings
of the Esteve Foundation Symposium IV, Son Vida, Mallorca, 7-10
October 1990 / editor, M.M. Reidenberg.
        p.   cm. -- (International congress series ; no. 944) (Esteve
Foundation symposia ; vol. 4)
    Includes indexes.
    ISBN 0-444-81376-4 (alk. paper)
    1. Proteins--Therapeutic use--Testing--Congresses. 2. Peptides-
-Therapeutic use--Testing--Congresses. 3. Biological products-
-Therapeutic use--Testing--Congresses. 4. Pharmaceutical
biotechnology--Technique--Congresses.   I. Reidenberg, Marcus M.,
1934-   . II. Title. III. Series. IV. Series: Esteve Foundation
symposia ; vol. 4.
    [DNLM: 1. Biological Products--pharmacology--congresses.
2. Biological Products--therapeutic use--congresses.
3. Biotechnology--congresses.   W3 EX89 no. 944 / QW 800 E79 1990c]
RM666.P87E77   1990
615'.3--dc20
DNLM/DLC
for Library of Congress   91-9498
                                                                CIP

Printed in The Netherlands

(handwritten library markings: Rm 301.5 E85 1990 Vet Med)

About the Esteve Foundation

The Esteve Foundation was established in 1983 to honor the late Dr. Antonio Esteve in the manner that would best fit his temperament, intellectual curiosity and dedication to science. Antonio Esteve was born in Manresa (Barcelona) in 1902, obtained a Doctorate in Pharmacy and in 1927 became manager of the pharmacy that his great-grandfather had founded in 1787. Having been actively involved in research during his student years. Dr. Antonio Esteve felt an urge for drug investigation that resulted in the establishment, in 1929, of what was to become one of the largest pharmaceutical firms in Spain. Dr. Antonio Esteve was actively involved in its research activities, actually heading this department, until his death in 1979.

The Esteve Foundation operates independently of any pharmaceutical enterprise and its main goal is to stimulate progress in pharmacotherapy through scientific communication and discussion. As a way to promote international cooperation in research, it organizes international multidisciplinary meetings – The Esteve Foundation Symposia – as well as meetings on topics of interest in its geographical area. The Esteve Foundation also sponsors lectures – among them an Antonio Esteve Lectures international series – seminars, courses and study groups on areas related to pharmacotherapy.

Introduction

The new techniques of molecular and cell biology have presented many opportunities for developing new agents for the diagnosis, treatment, and prevention of disease. Over 150 proteins and peptides are currently undergoing clinical investigation in the United States at this time. The number actually marketed goes up with increasing speed every year.

The excitement of moving the advances of biotechnology to the bedside increases with the increasing number of new compounds with therapeutic potential that become available. As the therapeutic potential becomes reality with one after another protein achieving registration and marketing, ideas emerge and questions related to issues about proteins and peptides as drugs are asked.

This symposium was planned to bring together researchers in the new biology, clinical pharmacologists, physicians, and administrators to share experiences and think together about how to convert the potential of modern biomedical science to the reality of improving the health for everybody. We have organized the symposium around several themes. The first theme is a general look at biotechnology as a source of drugs. Next, consideration will be given to the disposition and pharmacokinetics of peptides and proteins since these differ from the disposition and kinetics of small organic molecules, the traditional types of drugs. Assessing the safety of biotechnology products also has some difference from safety assessment of small organic molecules and is the next theme of the symposium. This is followed by some examples of pharmacodynamic and therapeutic effects of some of the medications derived from biotechnology. Finally, an overview of some of the ways the products of the new biology impact on society will be considered. These focus on cost and economic considerations since costs are of paramount importance for the utilization of biotechnology products in everyday medical practice.

By addressing issues of importance in the development and utilization of drugs derived from biotechnology, the organizers hope that this symposium will contribute to helping the new biology, through biotechnology, achieve its potential of improving the health of all people everywhere.

MARCUS M. REIDENBERG, M.D.
Departments of Pharmacology and Medicine,
Cornell University Medical College,
1300 York Avenue,
New York, NY 10021,
USA

List of participants

W. Aulitzky: Department of Urology, Landeskrankenanstalten Salzburg, Müllner Haupstrasse 48, A-5020 SALZBURG (Austria)

J. Bigorra: Departamento Médico, Química Farmacéutica Bayer, S.A., Calabria, 268, 08029 BARCELONA (Spain)

D.C. Brater: Clinical Pharmacology Section, Wishard Memorial Hospital, Indiana University School of Medicine, 1001 West 10th Street, WOP 320, INDIANAPOLIS, IN 46202 (U.S.A.)

C.A. Dionne: Rorer Biotechnology Inc., Department of Molecular Biology, 680 Allendale Road, King of Prussia, PA 19406 (U.S.A.)

S. Erill: Esteve Foundation, Llobet i Vall-Llosera, 2, 08032 BARCELONA (Spain)

J.A. Galloway: Lilly Research Laboratories, Lilly Corporate Center, INDIANAPOLIS, IN 46285 (U.S.A.)

A. Ganser: Department of Hematology, Johann Wolfgang Goethe-University, Theodor-Stern-Kai, 7, D-6000 FRANKFURT 70 (Germany)

F. García-Alonso: Ministerio de Sanidad y Consumo, P° del Prado, 18-20, 28014 MADRID (Spain)

L. Gauci: Department of Clinical Research, F. Hoffmann-La Roche, Ltd., CH-4002 BASEL (Switzerland)

A.J.H. Gearing: British Bio-technology Limited, Watington Road, Cowley, OXFORD OX4 5LY (England)

P. Graepel: Pharmaceuticals Division, Toxicology, Ciba-Geigy Limited, CH-4002 BASEL (Switzerland)

P. Juul: Department of Pharmacology, The Danish Royal School of Pharmacy, 2 Universitetsparken, DK-2100 COPENHAGEN Ø (Denmark)

M. Levy: Department of Medicine A, Clinical Pharmacology Unit, Hadassah University Hospital, JERUSALEM (Israel)

H.R. Lijnen: Center for Thrombosis and Vascular Research, Katholieke Universiteit Leuven, Herestraaat 49, B-3000 LEUVEN (Belgium)

D. Marhun: Bayer AG, Pharma Forschungszentrum, Fachbereich Medizin, Aprater Weg, Postfach 10 17 09, D-5600 WUPPERTAL 1 (Germany)

B.R. Meyer: Division of Clinical Pharmacology, North Shore University Hospital, Cornell University Medical College, 300 Community Drive, MANHASSET, N.Y. 11030 (U.S.A.)

A. Moreno: Servicio de Farmacología Clínica, Hospital Clínico de San Carlos, Ciudad Universitaria 28040 MADRID (Spain)

J. Mous: Central Research Units, F. Hoffmann La Roche, Ltd., CH-4002 BASEL (Switzerland)

M.M. Reidenberg: Division of Clinical Pharmacology, The New York Hospital-Cornell Medical Center, Box 70, 1300 York Avenue, NEW YORK, N.Y. 10021 (U.S.A.)

H. Ronneberger: Research Laboratories, Behringwerke AG, Postfach 1140, 3550 MARBURG/LAHN 1 (Germany)

R. Rivera: Hospital Division, Laboratorios Pensa, Torre dels Pardals, 29, 08026 BARCELONA (Spain)

H.J. Röthig: Department of Clinical Pharmacology, Hoechst AG, Postfach 80 03 20, 6230 FRANKFURT AM MAIN 80 (Germany)

J. Segura: Departamento de Farmacología y Toxicología, Institut Municipal d'Investigació Mèdica, Paseo Marítimo, 25-29, 08003 BARCELONA (Spain)

P. du Souich: Department of Pharmacology, Faculty of Medicine, University of Montreal, P.O. Box 6128 Station 'A', MONTREAL, P.Q. H3C 3J7 (Canada)

M. Steinmetz: Central Research Units, F. Hoffmann La Roche, Ltd., CH-4002 BASEL (Switzerland)

P. Tanswell: Department of Pharmacokinetics, Dr. Karl Thomae GmbH, Postfach 1755, Birkendorfer Strasse, 65, D-7950 BIBERACH AN DER RISS (Germany)

C. Tristán: Ministerio de Sanidad y Consumo, P° del Prado, 18-20, 28014 MADRID (Spain)

F. Valderrábano: Servicio de Nefrología, Hospital General Gregorio Marañón, Dr. Esquerdo, 46, 28007 MADRID (Spain)

W.M. Wardell: Department of Medicine, Boehringer Ingelheim Pharmaceuticals, Inc., 90 East Ridge, P.O. Box 368, RIDGEFIELD, CT 06877 (U.S.A.)

R.G. Werner: Department of Biotechnical Production, Dr. Karl Thomae GmbH, Postfach 1755, Birkendorfer Strasse, 65, D-7950 BIBERACH AN DER RISS (Germany)

Contents

xiv

BIOTECHNOLOGY AS A SOURCE OF DRUGS

PURIFICATION OF PROTEINS: CONVERTING A CULTURE BROTH INTO A MEDICINE

ROLF G. WERNER

Dr. Karl Thomae GmbH, Department Biotechnical Production, P.O. Box 1755,
7950 Biberach an der Riss, FRG

SUMMARY

In living organisms proteins maintain functions important to life. Faulty functioning or deficiency of proteins give rise to pathological reactions. Such physiologically occurring proteins can now be produced, using the methods of recombinant DNA technology and administered to patients for replacement therapy. A number of proteins as active ingredients of pharmaceutical preparations are already available for therapeutic use as immunomodulators, agents for tumour treatment, plasma proteins and hormones. They are in various stages of development, ranging from cloning of the producing cells to marketing of the finished products. Since the active substances are proteins synthesized by recombinant cells, their purification presents a particular challenge to protein chemists.

After fermentation of the appropriate production organism, hosting the gene for the protein synthesis, the purification of recombinant DNA-derived proteins intended for human use is an essential part of the biotechnical process. The characteristics of the protein determine if microorganisms or cell cultures are used and this in turn defines the first purification step. The microorganisms are disrupted, and the insoluble protein, deposited in insoluble inclusion bodies, has to be renatured, or the proteins secreted by mammalian cells have to be concentrated. The subsequent strategy for purification of the protein does not depend on the fermentation process but is entirely determined by the physicochemical properties of the proteins. The goal of the first purification step is to isolate as fast and quantiatively as possible the recombinant DNA derived protein from the culture filtrate, in order to minimize potential changes brought about by proteases or glycosidases. Immunoaffinity or ligand-affinity chromatography is used preferentially for this purpose.

The concentration of protein and buffer changes are carried out by precipitation followed by reconstitution or, preferably, by dialysis and ultra-filtration/diafiltration. The methods used for the separation of cellular proteins from recombinant DNA derived proteins are combinations of anion and cation exchange chromatography and gel permeation. Nucleic acids are removed efficiently by anion exchange chromatography on DEAE-columns or by degradation with nucleases.

Potential viral contaminants which might be associated with mammalian cell culture are preferably removed by ultrafiltration or inactivated by detergents. The purification process, which is carried out under aseptic conditions, is completed by sterile filtration, followed by freeze-drying thereby stabilizing the product. The efficiency of the purification process and the quality of the final product depend on the selection and sequence of the purification steps.

For therapeutic use in patients, active proteins must be free of microbial and viral contaminants. The content of protein impurities must be below 0.1%. According to international guidelines the limit for contamination by nucleic acids may not exceed 100 pg per dose. Biological contaminants, which are difficult to detect at low concentrations in the final product, must be removed by a validated method of processing, and the purification factors must be sufficiently effective.

1. INTRUDUCTION

Proteins are essential for the maintenance of bodily functions. Structural proteins contribute to the construction of cell membranes, intracellular microtubuli and microfilaments. The porins transport substrates and ions through the cell membrane, which controls the energy supply and osmotic pressure in the cell.

Enzymes catalyze chemical transformations taking place in the body, with substrates undergoing anabolism or catabolism to supply energy, detoxifi-cation, or metabolism for the de novo synthesis of essential compounds. Proteineous hormones regulate the body's metabolism and maintain the balance of synthetic activity. Protein receptors transmit signals which, in turn, initiate metabolic reactions. The antibodies are part of the immune response of the body, defending it against microbial or viral infections.

Generally, proteins act to maintain the state of physiological equilibrium in the body. Malfunctioning of these proteins can result in reactions which cause pathological states at the level of the protein molecules. This situation has led to the logical strategy of restoring the physiological equilibrium by providing the body with sufficient quantities of the proteins involved.

The low concentrations of functional proteins in human organs or body fluids, and the risk of transmission of viral infections, such as hepatitis and AIDS, or Creutzfeld-Jakob disease, make it either impossible or too risky to obtain proteins from human tissue or body fluids for therapeutic purposes. Chemical synthesis is no alternative because of the high molecular weight and the complex structure of most of the medically relevant proteins or glycoproteins. At present, chemical synthesis of proteins is possible only up to a molecular weight of about 15.000 Daltons, although not on an economically basis. The only possibility of producing such proteins or glycoproteins for therapeutic purposes is by biotechnology based on recombinant DNA technology. This paper deals exclusively with the purification and quality control of recombinant DNA-derived proteins which have been obtained by fermentation of microorganisms or cell cultures for use as active ingredient in pharmaceuticals.

2. PROTEINS AND GLYCOPROTEINS OF MEDICAL RELEVANCE

Basic medical research has identified a large number of proteins whose faulty activity in the body leads to disease symptoms. According to their profile of action these proteins can be classified as immunomodulators, antitumour agents, plasma proteins and hormones (1). Producing active proteins by biotechnological methods is a relatively new but promising technology, that potential protein-drugs are found in various stages of development.

Human insulin, interferon alpha, interferon beta, interferon gamma, human growth hormone, tissue plasminogen activator and erythropoietin are already commercially available. Tumour necrosis factor, interleukins, coagulation factor VIII, cardiactin, calcitonin, epidermal growth factor, relaxin, renin, oncostatin, protein A and other products are undergoing intensive clinical trials.

In addition, many more proteins are still in the stage of cloning, fermentation scale-up, purification or preclinical evaluation.

Thus, the production of biologically active proteins on the basis of re-combinant DNA technology presents a challenge to develop effective and economic processes for producing proteins of high purity and their cha-racterization (2, 3). All these proteins are used for life saving appli-cations where no other pharmaceuticals are available or not satisfactory in patient benefit.

3. OUTLINE OF THE PROCESSING AND PURIFICATION METHODS

The first step in the processing of proteins produced biotechnologically is determined by the producing organisms chosen. Proteins with relatively low molecular weights, few disulfide bridges and no glycosylation of relevance to the biological or immunological action, can be produced economically by microorganisms. In contrast, proteins with high molecular weights, numerous disulfide bridges and glycosylation which is relevant for the action, can be produced only by cell cultures, since only they have a synthetic apparatus which guarantees correct folding of the protein via the disulfide bridges, and glycosylation comparable to that in the natural glycoprotein.

In the case of microorganisms the protein is found in insoluble form as an inclusion body in the cell, in the case of cell cultures it is secreted in soluble form (4). The consequence is that, with microorganisms the first processing step is to disrupt the cells and to renature the protein which is in the form of an insoluble inclusion body. Although this results in enrichment of the recombinant protein, it also causes contamination with intracellular proteins and components of the medium, as well as incorrectly folded recombinant proteins. This problem can be alleviated to a certain extent, by the use of an appropriate expression system, if the proteins can be secreted into the periplasmic space of Escherichia coli, instead of being concentrated in inclusion bodies (5, 6). The goal of this first processing step is to obtain a homogeneous solution of the recombinant DNA derived protein.

In contrast to this problem associated with microbial fermentation, which adds to the difficulties of purifying the recombinant DNA derived proteins, the proteins secreted by cell cultures are present in their natural dis-solved form (7).

For the efficiency and economy of the separation of potential conta-minants it is of decisive importance that these contaminants are reduced to a minimum already in the cell culture medium. Measures, which significantly facilitate a proteinchemical purification due to reduced protein contaminants, are serum-free or preferentially protein-free cell culture media and an efficient removal of the production cells without cell lysis from the culture medium containing the active protein.

Because of the low concentration of the recombinant DNA derived protein in the cell culture filtrate, the first processing step in this case, after removal of the cells, consists of concentration of the protein, usually by affinity chromatography or ultrafiltration.

From this stage of recovery onwards, further purification of the recombi-nant DNA derived protein can be designed irrespective of the producing organism – microorganism or cell culture – and is chiefly determined by the physicochemical properties of the protein. The most important of these are: the molecular weight, the shape, the charge – determined by the amino acid composition and the glycosylation – and the hydrophobicity of the protein. These properties result in the following methods being used to purify the proteins: based on the shape and size of the molecule – ultra-filtration and gel permeation; based on the charge – ion exchange chromatography; based on the hydrophobicity – precipitation or affinity chromatography (8); and based on the specific affinities of active centres of the protein – chromatography using corresponding ligands or immobilized metal ions. Although this is a sufficient number of separation methods, the skill is to select the optimal combination of purification steps. The design of the purification process determines the yield, and thus the cost of the pro-cess, and often also the quality of the product.

In the first purification step as much as possible of the recombinant protein should be concentrated from the culture filtrate, in order to minimize any breakdown by proteases or glycosidases during the downstream processing. Monoclonal antibodies are particularly suitable since they are covalently bound to chromatography materials and recognize specific epitopes of the recombinant protein (9, 10). Thus, immunoaffinity chromatography is frequently used for the quantitative isolation of recombinant DNA derived proteins such as interferons, tumour necrosis factor or tissue plasminogen activator. However, despite the selective advantages of this method, it is relatively cumbersome: hybridoma techniques must be used to produce large amounts of monoclonal antibodies which, after appropriate purification, are then coupled to suitable chromatography materials, such as cyanogen bromide activated agarose. The useful lifespan of these immuno-affinity columns are relatively short. In addition, antibodies may become detached from the matrix during elution of the recombinant DNA derived protein. Moreover, because transformed cells have been used to produce the monoclonal antibodies, their freedom from oncogenic material must be demonstrated.

GMP-Guidelines for the pharmaceutical production of proteins for parenteral use require for all reagents which will be in contact with the product during the manufacturing process, that they have to be produced, analyzed and certified according to GMP-Guidelines. In accordance to these guidelines, the FDA requires for the production of monoclonal antibodies used as ligands for immunoaffinity chromatography that they are produced comparable to a therapeutic agent. This implies an expensive GMP-production for the monoclonal antibody itself in order to obtain the ligand. When using immunoaffinity chromatography, additional purification steps must be validated to remove any antibody contamination by column leakage, analytical assays must be established respectively.

Thus, a common alternative to immunoaffinity chromatography for the first purification step is bioaffinity chromatography. This method makes use of the specific binding of the protein, such as substrate specificity or specific affinity for metal ions (11, 12). The recombinant DNA derived protein is selectively purified by adsorption onto appropriate ligands coupled to the chromatography material. The ligands often used for this purpose are peptides or aminoacids which correspond to a specific centre of the enzyme. Thus, efficient purification of urokinase is possible with benzamidine-Sepharose (13) or pyroglutamyl-lysyl-leucyl-arginal-Agarose (14), and of tissue plasminogen activator with Lysine-Sepharose (15).

Basic matrices which allow an application of this technology at a technical scale are preferably Sepharose FF, Affi Gel or Eupergit because of their rigidity. Ligands, which specifically interact with the recombinant DNA derived protein, are linked to these matrices via appropriate spacers to avoid steric hindrance.

Affinity chromatography is specially designated for quantitative purification of proteins present in small concentrations in large volumes and with a high degree of contaminating proteins. The separating performance and capacity correlates with the specificity of the ligand. A rapid and economic purification of monoclonal antibodies for therapeutic use can be achieved by using affinity chromatography on protein A- or protein G-substituted matrices such as Repligen protein A-agarose or Genex Gammabind Plus.

For subsequent purification steps such as precipitation, followed by centrifugation or filtration, the protein solutions can be concentrated and the buffer be changed. Reconstitution of the precipitate results in transfer of the protein into a suitable buffer solution. However, recent developments in membrane technology have lead to an increase in the use of ultrafiltration and diafiltration techniques for concentrating and exchanging the buffer in protein solutions. In this method the protein remains in its natural, dissolved form (16, 17).

Electrical charge of proteins resulting from basic or acid amino acids are suitable for chromatographic separations on ion exchangers. This separation procedure is suited for the chromatography of large volumes during the initial procesing steps, also providing a good separating performance at a high capacity (up to 15 mg/ml gel) and a high linear flowrate (up to 1,500 cm/h). Chromatography materials preferably used in this aspect are ion exchange resins based on highly cross-linked agarose or equivalent matrices concerning rigidity and sanitization. There is no obvious preference for weak or strong ion exchangers as both types have their advantages: strong ion exchangers allow very robust process steps because of their relative insensitivity over a broad pH-range, which becomes important for elimination of potential hazardous contaminants, e.g. DNA removal by anion exchange chromatography. On the other hand, weak exchangers are more accessible to fine tuning of physical conditions regarding pH-value and ionic strength of the sample, but for this the corresponding purification step is somewhat sensitive to any deviation of physical conditions.

Combinations of anion and cation exchange chromatography are suitable for subsequent removal of cellular proteins or of proteins from the nutrient medium. A separation according to the molecular weight, by gel permeation, is also suitable for this purpose.

The hydrophobicity of proteins as a selective parameter can be exploited to full advantage on particularly substituted matrices of Sepharose, Fract-ogel TSK, Eupergit as well as CPG (Controlled Pore Glass) and Bioran. These procedures are appropriate primarily for materials with a high ionic strength following precipitation or ion exchange chromatography. Hydro-phobic matrices can be utilized in two modes: for hydrophobic interaction chromatography, using buffers of high and low salt molarities and for low pressure reversed phase chromatography, using buffers which contain organic solvents such as ethanol, 1-propanol or 2-propanol as an eluting agent. By this, the reversed phase mode predestinates such a purification step to be validated to inactivate potential viral contaminants.

Gelpermeation chromatography with high resolving matrices such as Sephacryl HR or Superdex allows separations according to molecular sizes of proteins. The efficiency of separation depends extremely on the volume of the sample to be separated, therefore gelpermeation is designated to be one of the latest steps in protein purification. Gelfiltration with conventional dextran gels is primarily suited for buffer exchange or formulation.

For economic reasons chromatographic materials for large scale purifica-tion of recombinant DNA derived proteins intended for therapeutic use must fulfill some basic requirements: high separating performance and capacity (> 10 mg product/ml gel) at a constant quality, pressure stability at high flow rates (> 200 cm/h), possibility of scaling-up from laboratory conditions to production scale, and long life time (> 100 cycles). In addition, the materials must be suited for sanitization and the initial costs should be low. In this respect the sanitization has for the production of bio-technically synthesized pharmaceuticals the goal to achieve sterility of the column materials by treatment with 0.1 - 1 M NaOH, 0.01 % Merthiolate/Thimerosal or 25 % ethanol. It is also used for the removal of pyrogens by 0.5 - 1.0 M NaOH and the elimination of ionically bound proteins by 2 M NaCl or that of hydrophobically bound proteins by 1 M NaOH, 3 M Na-Thiorodanite or organic solvents like 75% propanol. The materials used for chromatography must resist these sanitization procedures without impairment of separating performance or life time.

Since recombinant proteins of these types are used almost entirely for parenteral administration, processing and purification are carried out under aseptic conditions, with the buffer solutions and final product being sterilized by filtration and stored at low temperatures. These steps minimize the risk of microbial contamination or protein denaturation, and the pyrogen content of such parenteral products can be controlled.

In most cases, after downstream processing the recombinant DNA derived protein has been obtained in a purity greater than 99 % and is subjected to freeze-drying. The freeze-dried product is very stable over years, even at room temperature.

For economic reasons the design of the purification process is of decisive importance because the sequence and number of process steps determines yield, purity and costs of downstream processing. However, the selection of appropiate purification schemes is determined by physicochemical protein parameters and the particular contaminants (18).

4. GUIDELINES FOR THE PURIFICATION OF RECOMBINANT DNA-DERIVED PROTEINS

The production of recombinant DNA derived proteins, which are predominantly administered intravenously or subcutaneously, repeatedly or in high, non-physiological doses, must be reproducible and provide adequate purity (19, 20, 21, 22, 23, 24, 25). This is the only way to guarantee effects comparable with the natural product and to prevent immunological or toxicological reactions due to contaminants. Therefore, it is necessary to validate the purification process for removal of antigenic material originating from the producer cell or the culture medium. The content of protein impurities must be below 0.1%.

Concerning contamination with potentially infectious material, such as viruses or microorganisms which may be associated with the producer strain, it is necessary to demonstrate that these can be removed entirely by a number of purification steps. The final product must be free of viral and microbial contamination.

5. REMOVAL OF POTENTIAL HAZARDOUS CONTAMINANTS

5.1 Potential viral contaminants

Measures for the removal of potential viral contaminants can be taken all the more reliably, if only cell cultures essentially free of endogenous viruses are employed for the expression of recombinant DNA derived proteins. For the fermentation only batches of fetal calf serum tested for absence of viruses should be used. For the proteins which are added to the medium, validated purification schemes suited for the removal of potential viral contaminants should be employed.

In most cases of cell culture fermentation processes viral contaminants are not detectable. Therefore, spiking experiments in which distinct viruses are added the cell culture fluid must be performed for the validation of the purification procedure. As not all viruses can be ascertained and examined with these spiking experiments, a selection should cover all types of viruses if possible (26).

ß-Propiolactone, UV-light, gamma-irradiation, extreme pH-values, high temperatures or high concentrations of chaotropic salts such as urea or

guanidine are suited for virus inactivation. With the use of these methods
it must be ensured that the active protein is not influenced negatively by
the treatment. In addition, an efficient removal of viruses by means of
ultrafiltration through 300 KD membranes is feasible. The virus retention
with Sindbis virus being a 6 log reduction starting from a virus titer of
5.7×10^9 PFU (plaque forming units), the murine xenotropic virus, a 40 nm
retrovirus, could be retented to a full 4 log reduction starting from a
virus titer of $1 - 5 \times 10^4$ FFU (focus forming units).

During the fermentation process virus tests concerning endogenous and ad-
ventitious virus must be performed in order to guarantee that the validated
purification scheme eliminates potential contaminants. Purification factors
for the inactivation and removal of potential viral contaminants of individual elimination steps are in the range of $10^3 - 10^{10}$, depending on the
types of virus to be eliminated. A definite recovery process should contain
different steps which account to a 12 log reduction of retrovirus like the
murine xenotropic virus.

5.2 Contaminating nucleic acids

Based on tumorigenicity studies with intact viral nucleic acid with oncogenic potential the WHO (27) defined the maximum concentration of DNA for
reasons of product safety to 100 pg DNA per dose. This concentration remains under the tumour inducing dose by a factor of 10^{10}. Thus in respect
of drug safety the purification procedure must be validated accordingly.

The content of nucleic acid essentially is minimized by a high degree of
viable cells in the fermentation broth and by efficient removal of the cell
mass from the cell culture fluid without cell lysis and thus the proteinchemical purification is facilitated. In most cases the nucleic acid
concentration can be quantitatively determined only in the cell culture
medium or after the inital process step. Therefore, in particular spiking
experiments radioactive labelled DNA is added to the cell culture fluid in
such an amount that the desired purification factor can be quantitatively
established. Procedures of choice for nucleic acid removal are affinity
chromatography which selectively binds the active protein molecule leaving

nucleic acids in the flowthrough volume, or ion exchange chromatography using anion exchangers like DE52 or DEAE-Sepharose-FF, to which nucleic acids bind quantitatively under appropriate physical conditions. By combination of different steps purification factors up to 10^{14} can be achieved.

5.3 Pyrogens

Aseptic processing procedures including methods such as "cleaning-in-place and sterilization-in-place for devices, sanitization of column materials and the use of pyrogen-free buffers considerably contribute to a pyrogen-free final product. While buffer solutions can be depyrogenized very easily and also efficiently by means of ultrafiltration with an exclusion limit of 10KD, apart from the above described preventive measures affinity chromatography is the best choice for the selective binding of the active protein and thus for the removal of pyrogens. Ion exchange chromatography methods, deep filtration or alternatively filtration with nylon filter material have been used for the removal of pyrogens. The pyrogen content in the final product should be below 1 endotoxin unit per dose to prevent a pyrogenic effect.

6. VALIDATION OF PROTEIN RECOVERY

Separation procedures must be validated such that they are robust against variations of the fermentation process and do not negatively influence the native structure of the product. This emphasize primarily the shifting of the quantitave relationship of the active protein to potential contaminants in the cell culture medium and protein stability during processing, whereby capacity and life cycles for filters and chromatographic material must guarantee identical conditions for each use, exhibiting reproducibility of elution profiles and meeting of in-process-control specifications (28).

7. QUALITY CONTROL OF THE PURIFIED PRODUCT

The development of an efficient purification procedure is only possible if expressive protein-analytical methods for the determination of the purity, identity and activity of the product are available. Due to the complex structure of the proteins to be examined, numerous analytical methods must be applied whose results as a whole testify the quality of the product.

7.1 Determination of purity

For the determination of the purity the following methods are available:

SDS-PAGE for the determination of the relative molecular weight, detection of oligomeric structures and contaminants above 200 ppm of single contaminating proteins. Contaminants below this detection limit or possessing electrophoretic properties comparable to those of the active ingredient are lost for detection.

Isoelectric focussing as a guarantee for the constancy of complex proteins, for the determination of the microheterogenicity of glycoproteins, detection of deamidation of glutamine or asparagine.

HPLC-size exclusion chromatography for the determination of oligomeric structures and proteolytic cleavages.

Multiantigen-ELISA for the recovery of low concentrations (1 - 100 ppm) of contaminating proteins from the production cell or culture medium.

DNA-hybridization test for the determination of contaminations with nucleic acid with a detection limit in the nanogram range.

7.2 Determination of identity

For the determination of the identity the following protein-analytical methods are useful:

Amino acid analysis for smaller proteins (< 15,000 Dalton) and for the determination of norleucine substitution of methionine.

Partial sequencing of 8 - 10 amino acids at the N-terminal end of the protein and for the determination of potential proteolytic cleavages in this section.

Peptide mapping by HPLC separation of peptides following proteolytic cleavage of the active protein for the determination of the primary protein structure including the microheterogenicity of carbohydrate containing peptides, N- and C-terminal peptides, point mutation in the section of the cleavage sites and methionylated proteins with an additonal N-terminal amino acid methionin.

Immunoassay for epitope mapping of the active protein.

Carbohydrate analysis via determination of the neutral sugars, the sialic acid content and the sequence of monosaccharides as well as the enantiomeric configuration of glycoside links.

Circular dichroism for the determination of secondary and tertiary structures, conformation changes and structural equivalence.

7.3 Determination of activity

Analytical methods establishing the biological activity of the active protein are primarily determined by the mode of action of the protein. However, for the pursuit of process steps during purification and for the determination of the content of the active ingredient in the final product only those methods are appropriate which provide a respective accuracy. These are preferably enzymatic tests or biomimetic assays which allow the active principle to be reproduced (26).

8. FUTURE PROSPECTS FOR THE PRODUCTION OF RECOMBINANT DNA DERIVED PHARMACEUTICALS

Fundamental medical research, has helped to explain pathological states at the level of the protein molecule and molecular biology, has demonstrated the relationship between the nucleotide sequence in the genetic code and the amino acid sequence in the protein. This has provided the foundation for the new biotechnology, based on recombinant DNA technology, to produce endogenous proteins for replacement therapy in patients (29, 30). On the one hand, this is a challenge to protein chemists to prepare such recombinant DNA derived proteins of human origin in a pure state and to characterize them, and, on the other hand, it is an exceptional opportunity to provide new and promising approaches to treatment of patients. This applies particularly where chemically synthesized alternatives are not available or not satisfactory.

REFERENCES

1.   Werner, R.G. (1987) Biobusiness in pharmaceutical industry, Arzneimittel-Forschung/Drug Res. $\underline{37}$, 1086 - 1093

2.   Howell J., Krämer P., Nybergh P., van't Riet K., Schmidt-Kastner G., Skogman H., Sittig W., Voser W. (1986) Downstream processing in biotechnology. Recovery and purification of bioproducts. D. Behrens, DECHEMA, Frankfurt/Main (1) - 250

3.   Janson J.-Ch., Hedman P. (1982) Large-scale chromatography of proteins. Adv. Biochem. Eng. $\underline{25}$, 43 - 99

4.   Pestka S. (1983) The human interferons - from protein purification and sequence to cloning and expression in bacteria: before, between, and beyond. Archives of Biochemistry and Biophysics $\underline{221}$, 1, 1 - 37

5.   Hsiung H.M., Mayne N.G., Becker G.W. (1986) High-level expression, efficient secretion and folding of human growth hormone in Escherichia coli. Bio/Technology $\underline{4}$, 991 - 995

6.   Barbero J.L., Buesa J.M., Penalva M.A., Agustin P.-A., Garcia J.L. (1986) Secretion of mature human interferon alpha 2 into the periplasmic space of Escherichia coli. Journal of Biotechnology $\underline{4}$, 255 - 267

18

7. Cartwright T. (1987) Isolation and purification of products from animal cells. TIBTECH 5, 25 - 30

8. Gelsema W.J., Brandts P.M., de Ligny C.L., Theeuwes A.G.M., Roozen A.M.P. (1984) Hydrophobic interaction chromatography of aliphatic alcohols and carboxylic acids on octyl-sepharose CL-4B: mechanism and thermodynamics. Journal of Chromatography 295, 13 - 29

9. Chase H.A. (1983) Affinity separations utilising immobilised monoclonal antibodies - a new tool for the biochemical engineer. Chemical Engineering Science 39, 7/8, 1099 - 1125

10. Reagan M.E., Robb M., Bornstein I., Niday E.G. (1985) Immunoaffinity purification of tissue plasminogen activator from serum-supplemented conditioned media using monoclonal antibody. Thrombosis Research 40, 1 - 9

11. Sulkowski E. (1985) Purification of proteins by IMAC. Trends in Biotechnology 3, 1, 1 - 7

12. Rijken D.C., Collen D. (1981) Purification and charcterization of the plasminogen activator secreted by human melanoma cells in culture. The Journal of Biochemical Chemistry 256, 7035 - 7041

13. Winkler M.E., Blaber M., Bennett G.C., Holmes W., Vehar G.A. (1985) Purification and characterization of recombinant urokinase from Escherichia coli. Bio/Technology 3, 990 - 1000

14. Someno T., Saino T., Katoh K., Miyazaki H., Ishii S. (1985) Affinity chromatography of urokinase on an agarose derivative coupled with pyroglutamyl-lysyl-leucyl-Arginal. J. Biochem. 97, 1493 - 1500

15. Einarsson M., Brandt J., Kaplan L. (1985) Large-scale purification of human tissue-type plasminogen activator using monoclonal antibodies. Biochemica et Biophysica Acta 830, 1 - 10

16. Bio/Technology Staff Report: Filtration techniques for biotechnology productivity. (1986) Bio/Technology 4, 870 - 878

17. Tutunjian R.S. (1985) Scale-up considerations for membrane processes. Bio/Technology 3, 615 - 626

18. Werner, R.G. and Berthold, W. (1988) Purification of proteins produced by biotechnical process, Arzneim.-Forsch./Drug Res. 38, 422 - 428

19. Richtlinien des Rates vom 26.01.1965 (65/65/EWG), vom 20.05.1975 (75/318/EWG und 75/319/EWG) und vom 26.10.1983 (83/570/EWG).

20. Office of Biologics Research and Review Center for Drugs and Biologics. (1985) Points to Consider in the Production and Testing of New Drugs and Biologicals Produced by Recombinant DNA-Technology. Draft, April 10

21. Committee for proprietary medicinal products. Notes to Applicants for Marketing Authorization. On Requirements for the production and quality control of medicinal products derived by recombinant DNA-technology. Draft 7, March 1987

22. Renseignements a inclure dans le dossier de demande d'autorisation de mise sur le marche. Dossiers technique et analytique des produits biologiques a usage humain obtenus par les technologies utilisant recombinaison genetique ou anticorps monoclonaux. France 1985.

23. National Board of Health and Welfare. Department of Drugs. Pharmaceutical Division. General guidelines for registration of biosynthetic drugs based on recombinant DNA technique. Sweden 1983.

24. National Biological Standards Laboratory. Commonwealth. Deparment of Health. Guidelines for production of monoclonal antibodies intended for therapeutic use. November 1986.

25. Commission of the European Communities. Communication from the commission to the council. A community framework for the regulation of biotechnology. November 1986.

26. Werner R.G., Langlouis-Gau H. (1989) Meeting the Regulatory Requirements for Pharmaceutical Production of Recombinant DNA Derived Products. Arzneim.-Forsch./Drug Res. 39 (I), 1, 108 - 111

27. Report of a WHO Study Group (1987): Acceptability of all substrates for production of biologicals. Technical Report, Series 747, World Health Organization, Geneva

28. Werner R.G., Langlouis-Gau H., Walz F., Allgaier H. and Hoffmann H. (1988) Validation of biotechnical production process, Arzneim.-Forsch./Drug Res. 38, 855 - 862

29. Werner, R.G. (1986) New biotechnology as a logical outgrowth of molecular biology. Arzneim.-Forsch./Drug Res. 36, 9, 1429 - 1436

30. Werner R.G., Feuerer W. (1987) Neue Arzneimittel-Generation: Therapie mit körpereigenen Wirkstoffen. BioEngineering 4, 12 - 20

Discussion - PURIFICATION OF PROTEINS CONVERTING A CULTURE BROTH INTO
A MEDICINE

P. Juul

I have two questions. If you could choose between E. coli and yeast what would you prefer or, to put it into different words, what is the reason that different companies chose either E. coli or yeast? The second question concerns stability and storage problems. We accept a higher extent of degradation of some of the biological products. I am not afraid of a loss of potency of 10% but I'm afraid of the 10% which has gone to something which I don't know what it is, and might possibly constitute neoantigens.

R.G. Werner

Concerning the choice between E. coli and yeast, by using E. coli, one has a very cheap fermentation process, and the yield can be up to 3 g per liter. If one chooses the yeast, the fermentation costs are about the same, but the yield is far below, in the range of 500 - 800 mg/l. The advantage of choosing the yeast is that the molecule is secreted into the medium. It is there in its native form, so there is no need for refolding compared to E. coli. One has to take into account the cost of the recovery process. If one chooses E. coli for the production of an interferon molecule, for example, one needs, additional purification steps to separate uncorrectly folded molecules from the correctly folded molecules, and this is very difficult because the molecular weight and the charge of the molecule is identical. As far as the second question is concerned, an uncorrectly folded protein in a pharmaceutical preparation carries the risk that the immune response recognizes it as an antigen. If during the shelf life proteins are degradated and still used for therapy, antibodies against these fragments can be raised, which also can react against the endogenous protein.

A.J.H. Gearing

You said earlier that those molecules that you described were not too amenable to chemical synthesis. It's not actually quite true in that several of the cytokines, like interleukin 3, have been chemically-synthesized in gram quantities and seem to be fully active.

R.G. Werner

There is a possibility to chemical synthesis protein up to 15,000 molecular weight. Human growth hormone can be chemically synthesized. And, of course, insulin, with a molecular weight of 6000 was one of the first therapeutically useful proteins to be synthesized. However, the process is not economic because one loses too much at each step, even with a 99% yield in each chemical peptide bound synthesis.

HUMAN INSULIN AND ITS MODIFICATIONS

JOHN A. GALLOWAY, RONALD E. CHANCE, KENNETH S. E. SU

Lilly Research Laboratories, Eli Lilly and Company, Indianapolis, IN, USA

In the mid-1970s, a study by the National Diabetes Commission noted that the insulin-using diabetic population was growing faster than meat production and suggested the possibility of an animal pancreas shortage within the next 20-30 years[1]. In response, Eli Lilly and Company considered a number of alternative sources of insulin, including total chemical synthesis, tissue culture production, and production by genetic cell manipulation. After deciding upon the last[2], a project was undertaken with the University of California, San Francisco, and the Genentech Company to produce human insulin by recombinant DNA (HI rDNA) technology[2,3]. From the outset of the project, Lilly maintained a close working relationship with the Food and Drug Administration (FDA) and other regulatory agencies[2,4]. In anticipation of the New Drug Application (NDA) for HI rDNA, the Metabolic-Endocrine Division of the FDA recruited two senior scientists with the requisite expertise in biology and chemistry to deal with issues unique to recombinant technology[5]. To allay concerns about this novel manufacturing process, significant effort was made to educate both the scientific community and the public well in advance of producing human insulin for clinical trials[2].

The first dose of HI rDNA was administered to a normal volunteer in the U.K. in June 1980 and in the U.S. in August 1980, and to a diabetic patient in December 1980. An NDA was filed in the spring of 1982 with approval by the U.S. FDA the following fall. HI rDNA was the first drug for human use made by recombinant DNA technology and now accounts for 50% of insulin use in the U.S. and its use continues to grow.

Prior to the availability of human insulin prepared by recombinant DNA technology, limited quantities of material extracted from human pancreas had been given to humans[6-8]. Based on the similarities in chemical structure between human and pork insulin and on the results of these studies (Table I), it was anticipated that human insulin would be comparable to purified pork insulin (PPI) in pharmacology and multicenter clinical studies. However, as indicated below and summarized in Table II, certain differences were observed.

TABLE I
SPECIES DIFFERENCES IN AMINO ACID SEQUENCE OF MAMMALIAN INSULINS

	Positions			
	A Chain			B Chain
Source	8	9	10	30
Beef	Alanine	Serine	Valine	Alanine
Pork	Threonine	Serine	Isoleucine	Alanine
Human	Threonine	Serine	Isoleucine	Threonine

TABLE II
DIFFERENCES BETWEEN HUMAN AND ANIMAL INSULINS (PORK, MIXED BEEF-PORK) IN CLINICAL STUDIES

Clinical Pharmacology Studies:

Human insulin rDNA (neutral regular, isophane [NPH], and ultralente)
slightly more rapidly absorbed and shorter acting than purified pork insulin (PPI)[2,9,10]

Multicenter Studies in Insulin-Naive ("New ") Patients:

Metabolic control comparable to that of PPI and mixed beef-pork (MBP) insulin[11-13]

Serum insulin antibody titers slightly less in patients treated with human insulin rDNA than with PPI and significantly less than with MBP[13,14]

No differences in the frequency of allergy or lipoatrophy with HI rDNA, PPI, and MBP[13]

Studies in Established ("Transfer") Patients:

Slight increase in glycemia initially following transfer from PPI or MBP to human insulin[11,15]

A marked decrease in serum insulin antibody titers following transfer from MBP and minimal changes following transfer from PPI

No differences in the frequency of allergy or lipoatrophy[15]

Of interest is the fact that the findings generated from the use of HI rDNA have generally been confirmed by studies utilizing semisynthetic human insulin[10,16].

The fact that (as indicated in Table II) HI rDNA was found to be significantly less immunogenic than mixed beef-pork (MBP) insulin and slightly less immunogenic than PPI raised a question concerning the clinical significance of antibody binding. Clearly, binding >10%, a level more frequently reached in patients receiving PPI than HI rDNA[14,17] and in virtually all patients receiving beef-containing insulins, may result in an attenuation of the effect of injected insulin[18] and delay recovery from hypoglycemia[19] (Figure 1). Finally, because antibody binding promotes placental transfer of maternal insulin[20] the fetuses of mothers with increased serum insulin antibody titers are subject to hyperinsulinemia[21]. Thus, Menon *et al*[21] have shown that in pregnant patients with insulin-dependent diabetes mellitus (IDDM) increased antibody titers (binding >19.3 ± 2.3%) was associated with a significantly increased frequency of fetal macrosomia. Patients treated with HI rDNA had low mean antibody titers (5.0 ± 1%) and produced no infants with macrosomia.

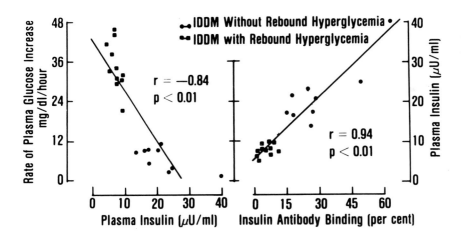

Fig.1. Effect of insulin antibodies on serum free insulin and recovery from hypoglycemia in patients with insulin-dependent diabetes mellitus. Reproduced with permission (ref. 19).

A final issue that arose from the use of HI rDNA (and semisynthetic human insulin) was the possibility that patients transferred from animal insulins (pork, beef, or a mixture) experienced fewer warning signs or symptoms of hypoglycemia[22]. Whether this phenomenon is real or apparent has been the subject of a number of studies and editorial comments[23]. The position of most authorities is that decreased awareness of hypoglycemia is more likely secondary to the diminution of cognitive function that results from "tight" glycemic control and is not related to insulin species[23,24].

Certainly, assurance of an unlimited supply of the homologous hormone which has advantages over animal-source insulin is a noteworthy scientific achievement. However, the most important therapeutic benefit for patients with diabetes mellitus to be derived from recombinant DNA technology may well prove to be the capability to design and produce analogues that are chemically and/or pharmacologically superior to native insulin. The impetus for this activity is based on the premise that failure to achieve the level of metabolic control deemed necessary to forestall the chronic complications of diabetes is in part due to the failure of conventional insulins to simulate the normal pattern of secretion of insulin in response to feeding and fasting. For example, as displayed in Figure 2, normal insulin secretion is characterized by (1) rapid and prompt increases in hormone levels to dispose of meals (peripheral glucose disposal [PGD]) followed by a correspondingly prompt return to basal levels and (2) low hormone levels in the postabsorptive, or basal state; these maintain normal glycemia between meals by reducing hepatic glucose production (HGP). The importance of the latter in the overall glycemic control of patients with diabetes is that the fasting blood glucose is highly correlated with HGP. Moreover, excessive HGP is a hallmark of noninsulin-dependent diabetes mellitus (NIDDM)[25].

Because studies with pork proinsulin had disclosed relative hepatospecificity, e.g., in comparison with insulin greater suppression of HGP versus promotion of PGD, proinsulin was viewed as a basal analogue which might be useful in suppressing HGP with a minimal risk of hypoglycemia. Moreover, pork proinsulin, which in contrast to the commercially available modified insulins that are suspensions of zinc and/or protamine, was a soluble intermediate-acting insulin agonist. Therefore, the technology that produced human insulin was applied to the production of the first biosynthetic analogue of insulin, human proinsulin (HPI), clinical testing of which began in mid-1982. The pharmacologic features of pork proinsulin were confirmed with HPI. Moreover, large intrasubject coefficient of variation of responses that are characteristic of treatment with insulin[26] were remarkably reduced by HPI[25]. Unfortunately, clinical studies of this agent were suspended due to an increased frequency of acute myocardial infarction in one multicenter study and the fact that in multicenter trials glycemic control was not different in HPI-treated versus insulin-treated groups (i.e., the attractive pharmacologic features apparently were not translated into clinical benefit)[25,27,28].

Jorgensen *et al* have reported on an insulin analogue that is substantially more slowly absorbed than ultralente beef insulin, the slowest acting of the commercially available modified insulins[29]. However, glucodynamic data on this analogue are not available.

Both Novo-Nordisk[30] and Lilly have utilized recombinant DNA technology to prepare insulin analogues which promote PGD more efficiently than insulin. The need for such analogues can be seen by examining Figures 2-4.

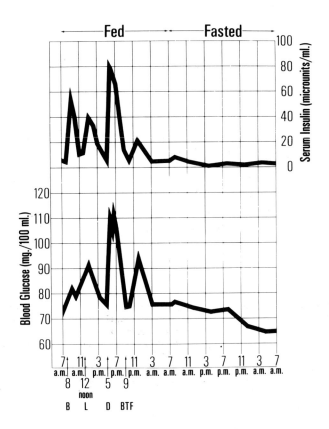

Fig. 2. The blood glucose and serum insulin response of six normal volunteers to fasting and feeding. Diet consisted of 30 calories per kg of body weight, 2/7 at each main meal and 1/7 at bedtime.

28

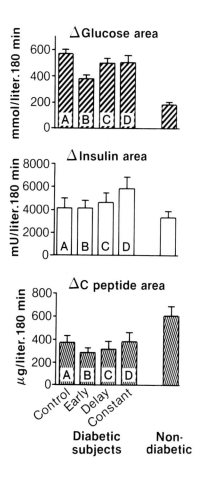

Fig. 3. The integrated glycemic, serum insulin, and C-peptide response of slightly overweight persons with NIDDM to a mixed meal containing 68 grams of carbohydrate to (A) no exogenous insulin; (B) a ramped intravenous infusion of 1.8 U of insulin administered between 0 and 30 min before the meal to simulate normal endogenous insulin secretion; (C) the same as (B) but the insulin infusion was begun with the meal; and (D) 1.8 U of insulin infused over 180 min. This figure demonstrates that the most efficient treatment (low blood glucose and serum insulin) was treatment (B). Reproduced with permission from Diabetes (1988) 37: 736-44.

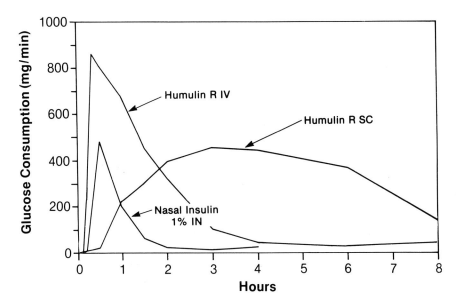

Fig. 4. The results of a glucose clamp (Biostator) study demonstrating the pharmacodynamic response of normal volunteers to human insulin rDNA 0.2 U/kg administered intravenously or subcutaneously or 1.4 U/kg applied to the nasal mucosa. Courtesy of D. C. Howey and S. A. Hooper.

Figure 2 shows that the prompt rise of serum hormone levels is useful in optimizing control of meal-induced hyperglycemia. Figure 3 demonstrates the increased effective-ness of insulin when given slightly before a meal. However, Figure 4 shows that although intravenous and nasally applied insulin simulate a serum insulin profile which is comparable to a normal response to a meal, because of a marked delay in peak after subcutaneous injection, regular or soluble insulin clearly does not. Moreover, the latter has an extended duration of action, features that can contribute to hypoglycemia between meals. Another disadvantage of subcutaneous regular insulin is that prolonged hyperinsulinemia has been implicated in the development of atherosclerosis and coronary heart disease[31]. The long-acting nature of soluble insulin has been shown to be due to its tendency to self-associate into dimers, tetramers, and hexamers. Preservation of the peptide in the monomeric form significantly speeds its absorption and disposal[30]. While it is beyond the scope of this paper to describe the chemistry of self-association of insulin, it is appropriate to point out that recombinant DNA technology provides a relatively convenient technique for the substitution of amino acids in the insulin molecule to minimize monomer-monomer interactions. Fortunately, the chemical goal of monomer preservation usually can be achieved with changing fewer than four, usually one or two,

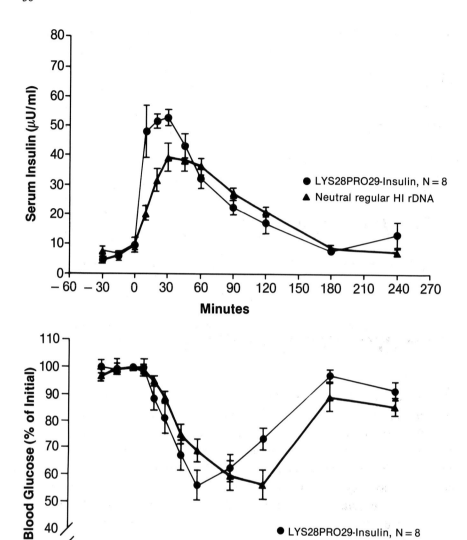

Fig. 5. The serum insulin and blood glucose response of anesthetized normal fasted male beagle dogs to the administration of neutral regular HI rDNA or the "Lys-Pro" analogue 0.1 U/kg subcutaneously. (The antibody used in the radioimmunoassay binds the two insulin agonists equally.)

of the 51 amino acids in the human insulin molecule. However, the task is complicated by the fact that the resulting new entity must (1) have the capacity of native insulin to interact with the insulin receptor and/or a slower metabolic clearance rate so that its net hypoglycemic potency is comparable to that of insulin, (2) be chemically stable, and (3) not be immunogenic. If the last condition is not met, antibodies might be formed which would bind the analogue, thereby defeating the pharmacodynamic purpose for which it was created[30]. An example of a quick-acting "monomeric" insulin is Lilly compound 275585 in which the proline and lysine at positions B28 and B29 of human insulin have been reversed to produce "Lys-Pro" insulin[32]. As shown in Figure 5, the Tmax is significantly higher and occurs sooner with the analogue than with HI rDNA. Moreover, serum concentrations of the "Lys-Pro" analogue dissipate more quickly than with HI rDNA. Studies are in progress to confirm the clinical utility of this and other monomeric analogues.

Recombinant DNA technology has also produced naturally occurring peptides that can be used as research probes. For instance, the availability of biosynthetic human C-peptide has made possible the detailed investigation of endogenous insulin secretion in health and disease by a noninvasive technique[33-37]. In addition, insulin-like growth factors (IGF-I and II) have been cloned and their interactions with other hormone receptors studied[38]. Metabolic studies in normal volunteers[39] have demonstrated that IGF-I has a hypoglycemic potency comparable to that of HPI and suggested an insulin-sparing effect. A recent report has suggested that IGF-I reduces insulin resistance in the muscle of normal-weight but not obese patients with NIDDM[40].

Overall, recombinant DNA technology has precluded a worldwide insulin shortage and provided means for producing a number of insulin agonists all of which will be useful research tools and some of which may well be found to be superior to existing insulin preparations in the treatment of diabetes mellitus. Clearly, the acceptance and success of the recombinant human insulin project has had a major positive impact on all drug development.

REFERENCES

1. A Study of Insulin Supply and Demand. A Report of the National Diabetes Advisory Board, Publication No. 78-1588 (1978). Washington, DC, US Dept of Health, Education, and Welfare

2. Galloway JA, Chance RE (1984) In: Lemberger L, Reidenberg MM (eds), Proceedings of the Second World Conference on Clinical Pharmacology and Therapeutics (Washington, DC, July 31-August 5, 1983). American Society for Pharmacology and Experimental Therapeutics, Bethesda MD, pp 503-20

3. Goeddel DV, Kleid DG, Bolivar F, Heyneker HL, Yansura DG, Crea R, Hirose T, Kraszewski A, Itakura K, Riggs AD (1979) Proc Natl Acad Sci (USA) 76:106-10

4. Johnson IS (1982) Diabetes Care 5(Suppl.2):4-12

5. Sobel S Personal Communication

6. Deckert T, Andersen OO, Grundahl E, Kerp L (1972) Diabetologia 8:358-61

7. Akre PR, Kirtley WR, Galloway JA (1964) Diabetes 13:135-43

8. Boshell BR, Barrett JC, Wilensky AS, Patton TB (1964) Diabetes 13:144-52

9. Galloway JA, Root MA, Bergstrom R, Spradlin CT, Howey DC, Fineberg SE, Jackson RL (1982) Diabetes Care 5(Suppl.2)13-22

10. Owens DR (1986) Human Insulin, Clinical Pharmacologic Studies in Normal Man. Falcon House, Lancaster, England, MTP Press Limited

11. Galloway JA, Peck FB Jr, Fineberg SE, Spradlin CT, Marsden JH, Allemenos D, Ingulli-Fattic J (1984) Diabetes Care 5(Suppl.2):135-39

12. Galloway JA (1985) Netherlands J Med 28(Suppl.1):37-42

13. Study IBAC data on file

14. Fineberg SE, Galloway JA, Fineberg NS, Rathbun MJ, Hufferd S (1983) Diabetologia 25:465-69

15. Study IBAL data on file

16. Ebihara A, Kondo K, Ohashi K, Kosaka K, Kuzuya T, Matsuda A (1983) Diabetes Care 6(Suppl.1):17-22

17. Fineberg SE, Galloway JA, Fineberg NS, Goldman J (1983) Diabetes 32:592-99

18. Van Haeften TW, Heiling VJ, Gerich JE (1987) Diabetes 36:305-09

19. Bolli GB, Dimitriadis GD, Pehling GB, Baker BA, Haymond MW, Cryer PE, Gerich JE (1984) N Engl J Med 310:1706-11

20. Bauman WA, Yalow RS (1981) Proc Natl Acad Sci (USA) 78:4588-90

21. Menon RK, Cohen RM, Sperling MA, Cutfield WS, Mimouni F, Khoury JC (1990) N Engl J Med 323:309-15

22. Teuscher A, Berger WG (1987) Lancet pp. 382-85

23. Gale EAM (1989) Lancet pp. 1264-66

24. Committee on Safety of Medicines (1990) Current Problems, Number 29 (August)

25. Galloway JA (1990) Diabetes Care (to appear in December issue)

26. Galloway JA, Spradlin CT, Howey DC, Dupre J (1986) In: Serrano-Rios M, Lefebvre PJ (eds) Diabetes 1985. Excerpta Medica, Amsterdam, pp 877-86

27. Galloway JA, Anderson JH, Spradlin CT (1989) In: Larkins RG, Zimmet PZ, Chisholm DJ (eds) Diabetes 1988. Excerpta Medica, Amsterdam, pp 85-88

28. Spradlin CT, Galloway JA, Anderson JH (1990) Diabetologia 33:A60

29. Jorgensen S, Vaag A, Langkjaer L, Hougaard P, Markussen J (1989) Br Med J 299:415-19

30. Brange J, Owens DR, Kang S, Volund A (1990) Diabetes Care 13:923-54

31. Stout RW (1990) Diabetes Care 13:631-54

32. European Patent Publication 383472, Published February 6, 1990.

33. Polonsky KS, Licinio-Paixao J, Given BD, Pugh W, Rue P, Galloway J, Karrison T, Frank B (1986) J Clin Invest 77:98-105

34. Shapiro ET, Tillil H, Miller MA, Frank BH, Galloway JA, Rubenstein AH, Polonsky KS (1987) Diabetes 36:1365-71

35. Polonsky KS, Given BD, Hirsch LJ, Tillil H, Shapiro ET, Beebe C, Frank BH, Galloway JA, Van Cauter E (1988) New Engl J Med 318:1231-39

36. Polonsky KS, Given BD, Hirsch L, Shapiro ET, Tillil H, Beebe C, Galloway JA, Frank BH, Karrison T, Van Cauter E (1988) J Clin Invest 81:435-41

37. Tillil H, Shapiro T, Given BD, Rue P, Rubenstein AH, Galloway JA, Polonsky KS (1988) Diabetes 37:195-201

38. Humbel RE (1990) Eur J Biochem 190:445-62

34

39. Guler HP, Zapf J, Froesch ER (1987) N Engl J Med 317:137-40

40. Dohm GL, Elton CW, Raju MS, Mooney ND, DiMarchi R, Pories WJ, Flickinger EG, Atkinson SM, Caro JF (1990) Diabetes 39:1028-32

Discussion - HUMAN INSULIN AND ITS MODIFICATIONS

D. Maruhn

I was somewhat surprised that you showed only results on glucose and no results on long term diabetic control, for example using HBA 1c or something like that.

J.A. Galloway

In the transfer studies, where the blood sugars were the highest, there was no statistically significant difference between the glycohemoglobin levels. The insulin dose was the same for both groups.

R.G. Werner

The insulin derivatives that you mentioned may have a great advantage because of their mode of action, but they might be immunogenic. Do you see any possibility to clear out the risk of immunogenicity prior to clinical trials?

J.A. Galloway

No, we are aware of no way to assess immunogenicity pre-clinically.

THE USE OF RECOMBINANT PROTEINS IN AIDS RESEARCH: DEVELOPMENT OF A CD4/GP120 BINDING ASSAY

JAN MOUS, CHRISTIAN MANZONI and LUC DIRCKX

Central Research Units, F.Hoffmann-La Roche Ltd, CH-4002 Basel, Switzerland

INTRODUCTION

The continuous increase in the number of people infected with the Human Immunodeficiency Virus (HIV), the causative agent of the acquired immunodeficiency syndrome, AIDS, mandates the development of effective strategies to stop the threatening pandemic. Recombinant DNA technology is of crucial importance not only to increase our knowledge of the biology of the AIDS virus, but also to allow the rapid development of accurate diagnostic tools, of new therapeutic approaches and protective vaccines. The availability of large amounts of recombinant proteins opens new perspectives in all areas of applied AIDS research, as illustrated by the following example.

The first step in the HIV infection cycle involves the binding of the viral envelope glycoprotein, gp120, to the cell-surface receptor, CD4, present on helper T-lymphocytes and monocyte/macrophages (1). Because of the high affinity interaction between gp120 and CD4 is required for infection with all strains of HIV, various therapeutic approaches based on this interaction have been proposed. One strategy has been to identify anti-CD4 mAb that mimic the binding site on gp120 for CD4 and use these as idiotypic vaccines (2-4). Another has been to identify the gp120 binding site on CD4 and design synthetic peptides which mimic the binding domain (5). A third, more promising approach has been to express soluble forms of the CD4 protein (6-10) or molecular fusions of CD4 with either toxins (11,12) or immunoglobulins (13,14) that block HIV infection and selectively kill HIV-infected cells. Although this strategy represents a potential antiviral therapy for HIV infection, it requires large amounts of recombinant protein to treat one patient; this may cause manufacturing problems. In addition, the therapy of AIDS patients is expected to be longlasting. The injection of large doses of recombinant soluble CD4 over extended periods could induce antibodies that neutralize the antiviral effect.

As an alternative approach we designed a gp120/CD4 binding assay using recombinant proteins. This system allows high-flux screening for low molecular weight, preferentially oral-bioavailable compounds that interfere with the binding of HIV to the target cell.

MATERIAL AND METHODS

Plasmid constructs and microbiological manipulations

A BamHI-HindIII cDNA fragment encoding the first two domains (V1+V2) of human CD4 was isolated from plasmid pHCD4 (kindly provided by A. Traunecker, Institute for Immunology, Basel) by introducing a BamHI restriction site at position aa+1 after the signal peptide cleavage site by site directed mutagenesis and by ligating a HindIII recognition sequence to the NheI site immediately upstream of the V2 splice junction. The resulting DNA fragment was then inserted into the E.coli expression vector pDS56/RBSII,6xHis (15). The CD4 C-terminal deletion mutants were made by Bal 31 exonuclease digestion from the HindIII site, followed by Klenow polymerase treatment and ligation of a HindIII linker. Expression of all recombinant CD4 proteins was in the E.coli strain M15 harbouring the lac repressor-producing plasmid pDMI,1 as described previously (16).

Purification of recombinant CD4

CD4, V1+V2 and all deletion mutants were expressed with a hexa-histidine affinity label at the amino terminus. This permits a rapid purification of mg quantities from a crude lysate by metal chelate affinity chromatography (15,17). Cultures of the recombinant bacteria were grown in antibiotic-supplemented L-broth until A_{600} ~0.7, after which ß-D-thiogalactopyranoside was added to a final concentration of 400 µg/ml. After an incubation period of 4h, the bacteria were collected by centrifugation and the pellets lysed in 6M guanidine hydrochloride, pH 8.0. The purification of 6His-CD4 over nickel nitrilotriacetate-Sepharose was then essentially done as described by Stüber et al. (15). CD4 bound to the column was eluted at pH6, renatured by dialysis against 10 mM Na-acetate pH 6.5 and stored frozen at - 80°C. By this method 75% pure, soluble CD4 could be recovered in one step from a crude bacterial lysate with a yield of 20-30 mg/l.

Construction of recombinant baculoviruses

The baculovirus transfer vector pVL941, obtained from M. Summers, Department of Entomology, Texas A&M University, was used to express HIV-1 gp120. From the HIV-1 proviral clone HAN2 (18) a AvaII to BsmI restriction fragment encoding gp120 was tagged with BamHI linkers and inserted into the BamHI site of pVL941 to generate pVL-gp120. Cotransfection of this plasmid together with wild type Autographa californica nuclear polyhedrosis virus DNA into Spodoptera frugiperda (Sf9) cells was by lipofection (Gibco). Recombinant occlusion negative viruses were isolated by limiting dilution followed by plaque purification on agar plates (19). Recombinant gp120 secreted into the culture medium of 200 ml spinner cultures of Sf9 cells infected with recombinant baculovirus was purified as described previously (20,21).

Immunological assays

The binding affinity of recombinant CD4 to the anti-CD4 mAbs Leu-3a (Becton-Dickinson) and OKT4a (Ortho), and the interaction of recombinant gp120 with recombinant CD4 were assayed by ELISA essentially as described by Gallati et al. (22).

Virological assay

The antiviral activity of the different recombinant proteins was measured in a syncytium formation assay. In short, a pretitrated amount of HIV-1, HAN (18) was incubated at room temperature with serially diluted recombinant proteins for 30 min. Thereafter, 25 µl of each mixture was transferred in triplicate into the wells of a 96-well plate which contained 25,000 MT-2 cells (23) per well in 50 µl of medium. After 3 days of culture, 100 µl of fresh medium was added and after 5 days the syncytia were counted in each well.

RESULTS AND DISCUSSION

Expression of soluble CD4 in E.coli

As firstly demonstrated by Traunecker et al. (10) the immunoglobulin-like and second domains of human CD4 are sufficient for binding the HIV envelope protein gp120 and to protect target cells against HIV infection. Therefore we introduced a cDNA fragment encoding the first two domains (V1+V2) into the E.coli expression vector pDS56/RBSII,6xHis (15) which directs the production of heterologous proteins N-terminally tagged with a hexa-histidine tail. This strategy allows the rapid purification of recombinant proteins by affinity

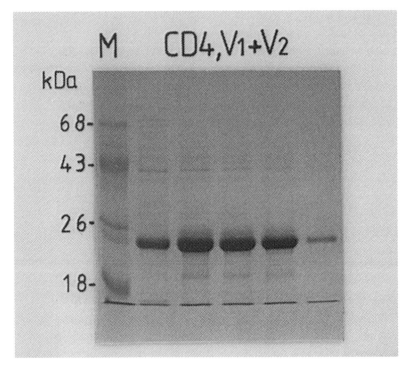

Fig. 1.　　SDS-PAGE analysis of recombinant CD4 present in the peak fractions eluted at pH6.0 from the NTA-column (see Materials and Methods). Proteins are visualised by Coommassie Blue staining. Lane M represents molecular weight standard proteins.

chromatography over a Ni-chelate column as described in Material and Methods. As shown in Figure 1 the recombinant CD4,V1+V2 eluted from the column was around 75% pure after a single purification step from the crude bacterial lysate. Next, the peak fractions of the eluate were pooled and directly dialysed against 10 mM Na-acetate, pH6.5. This simple method was found to give the highest yields of soluble CD4. Moreover, this CD4 preparation bound with high affinity to the mAbs Leu-3a and OKT4a, which recognize part of the gp120 binding domain on native CD4 (24), when measured in an ELISA-type assay (data not shown). This result demonstrates that our recombinant CD4,V1+V2 expressed in E.coli adopted the proper configuration to generate biological activity (which will be described later) in spite of the presence of artificial amino acids at the N- and C-terminus (see Fig.2).

Serial C-terminal deletions of CD4, V1+V2

Affinity purified soluble CD4 containing the domains V1 and V2 was tested for antiviral activity in a HIV-1 dependent syncytium formation assay using MT-2 cells

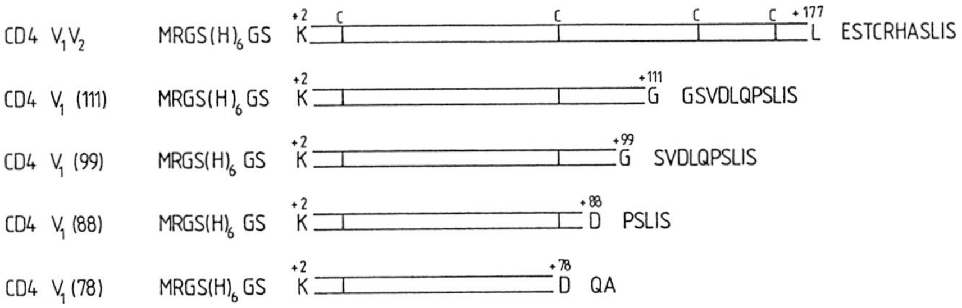

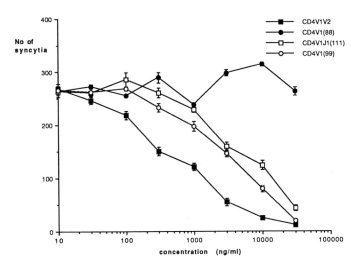

CD4 V_1V_2 MRGS(H)$_6$ GS   K ══════════════ L ESTCRHASLIS

CD4 V_1 (111)   MRGS(H)$_6$ GS   K ══════════ G GSVDLQPSLIS

CD4 V_1 (99)   MRGS(H)$_6$ GS   K ══════════ G SVDLQPSLIS

CD4 V_1 (88)   MRGS(H)$_6$ GS   K ════════ D PSLIS

CD4 V_1 (78)   MRGS(H)$_6$ GS   K ══════ D QA

Fig. 2.   Primary sequence of the different CD4 variants expressed in E.coli. One letter amino acid abbreviations are used.

Fig. 3.   Analysis of the antiviral activity of different recombinant CD4 molecules in a HIV-1 syncytium assay (see Materials and Methods). The indicated values are means of triplicate assays.

as outlined in the Material and Methods section. HIV-1 infection of MT-2 cells resulted in the formation of typical balloon-type cells (syncytia), on average 260 syncytia per well of a microtiter plate after 5 days of incubation. As illustrated in Figure 3, CD4,V1+V2 inhibited the occurrence of syncytia by 50% at a concentration of 0.8 µg/ml (IC_{50} = 0.8 µg/ml), which approximates the results published in the literature (6-9).

In order to determine the minimal length of soluble CD4 required for the observed antiviral effect, we made serial 3'-terminal deletions of the CD4 DNA insert by Bal31 exonuclease treatment as indicated in the Methods section. The expected amino acid sequences of the different CD4 proteins expressed from the mutant gene fragments is depicted in Figure 2. Note that the various CD4 mutants contain the same artificial N-terminal sequence, including the hexa-His label to allow a standard affinity purification over a Ni-chelate column. The primary sequence of the CD4 protein was as described by Mizukami et al. (25). After purification of the different recombinant CD4 variants to the same extent as described for CD4,V1+V2, the protein preparations were tested for their capacity to block HIV infection. The results of these experiments are shown in Figure 3. Both CD4 mutants truncated at position 111 and 99 of the primary sequence still retained the potency to inhibit HIV-infection. However, the mutant V1 (88) containing CD4 sequences up to amino acid position 88 completely lost the antiviral activity. The same result was obtained with mutant V1 (78) (data not shown). Concomitant with the loss of antiviral activity, mutants V1 (88) and V1 (78) also lost the reactivity with the mAbs Leu-3a and OKT4a.

Production of recombinant gp120 in insect cells

Relatively large amounts of the HIV-1 envelope glycoprotein gp120 can be produced when using the baculovirus expression system (20). Sf9 cells infected with recombinant baculovirus containing the gene for HIV-1,HAN gp120 (18) under the control of the polyhedrin promoter secrete ± 5 µg/ml gp120 into the culture medium at day 4 after infection (L. Dirckx, unpublished observation).

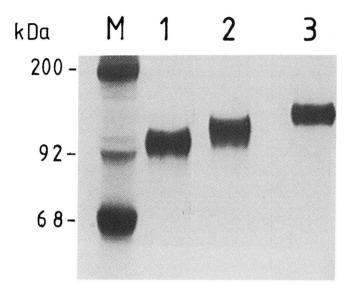

Fig. 4.　　Gel electrophoretic analysis of envelope glycoprotein gp120 immunoprecipitated from the culture medium (lane 1) or the cell lysate (lane 2) of Sf9 cells infected with recombinant baculovirus, and, for comparison, from the medium of HIV-1 infected JURKAT cells. Immunoprecipitations with a HIV+-serum were done as described previously (20).

As can be seen in Figure 4, the HIV envelope protein produced in insect cells is somewhat smaller than authentic gp120 produced in HIV-1 infected human T-cells. This is due to differences in glycosylation (20). Nevertheless, the recombinant gp120 was found to have the same high affinity ($Kd{\sim}3.10^{-9}M$) for the CD4 receptor as reported for the native envelope protein (21).

A prototype CD4/gp120 binding assay

The availability of mg amounts of both ligand (gp120) and receptor (CD4) allowed us to establish a screening assay for compounds that can interfere with the CD4/gp120 interaction. CD4,V1+V2 was purified from a bacterial lysate as described in Materials and Methods. Recombinant gp120 was isolated from a 200 ml culture of infected Sf9 cells by CD4-affinity purification (21). After coating microtiter plates with 10 µg/ml CD4,V1+V2 in 0.1M sodium hydrogen carbonate, pH8.0 (22), the different wells were incubated with increasing concentrations of purified gp120 in sample dilution buffer (22) for 1h at 37°C. After 2 washing steps with PBS/0.05% Tween-20, a 1:1000 dilution of HIV+-serum in PBS/Tween/20% FCS was added and incubated overnight at 4°C. Finally, extensive washing with PBS/Tween was followed by a 1h incubation at room temperature in the presence

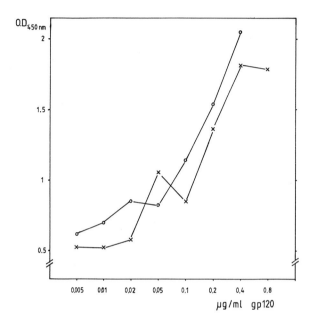

Fig. 5.   Titration of recombinant gp120 on CD4,V1+V2-coated microtiter plates by ELISA as described in the text. O-O and X-X represent two different batches of purified gp120.

of conjugate (goat anti-human IgG-peroxidase conjugate, 1:1000). Peroxidase activity was then measured by addition of 5.5'-tetramethylbenzidine (22). The results of these experiments are illustrated in Figure 5. As can be seen, recombinant gp120 interacted with CD4,V1+V2 in a dose-dependent way, showing saturation at a concentration of approximately 0.8 µg/ml. Therefore, a concentration of 1 µg/ml was used in our prototype CD4/gp120 binding assay, which can be used to identify novel compounds that may block the high affinity interaction of the HIV virus with its taret cells. Further improvements of this basic screening system are underway. The final goal is a high-flux, direct binding assay using recombinant CD4,V1+V2 and peroxidase-conjugated gp120 in a single step enzymatic immunoassay.

ACKNOWLEDGEMENTS

We like to thank R. Ette, M. Hänggi, S. Noack and K. Yasargil for their expert technical assistance. We also acknowledge the help of Dr. E.-J. Schlaeger, Dr. H. Döbeli and D. Moritz in the production and purification of recombinant gp120. We thank Dr. H. Gallati for his advice in the immunological assays and A. Traunecker for the CD4 cDNA clone. C.M. is a recipient of grant n° 1944.2 from the Swiss "Kommission zur Förderung der wissenschaftlichen Forschung".

REFERENCES

1. Sattentau QJ, Weiss RA (1988) Cell 52:631-633
2. McDougal JS, Nicholson JKA, Cross GD, Cort SP, Kennedy MS, Mawle AC (1986) J Immunol 137:2937-2944
3. Chanh TC, Dreesman GR, Kennedy RC (1987) Proc Natl Acad Sci USA 84:3891-3895
4. Dalgleish AG, Chanh TC, Thomson BJ, Malkovsky M, Kennedy RC (1987) Lancet ii:1047-1050
5. Lifson JD, Hwang KM, Nara PL, Fraser B, Padgett M, Dunlop NM, Eiden LE (1988) Science 241:712-716
6. Smith DH, Byrn RA, Marsters SA, Gregory T, Groopman JE, Capon DJ (1987) Science 238:1704-1707
7. Fisher RA, Bertonis JM, Meier W, Johnson VA, Costopoulos DS, Liu T, Tizard R, Walker BD, Hirsch MS, Schooley RT, Flavell RA (1988) Nature 331:76-78
8. Hussey RE, Richardson NE, Kowalski M, Brown NR, Chang HC, Siliciano RF, Dorfman T, Walker B, Sodroski J, Reinherz EL (1988) Nature 331:78-81
9. Deen KC, McDougal JS, Inacker R, Folena-Wasserman G, Arthos J, Rosenberg J, Maddon PJ, Axel R, Sweet RW (1988) Nature 331:82-84
10. Traunecker A, Lüke W, Karjalainen K (1988) Nature 331:84-86
11. Chandhary VK, Mizukami T, Fuerst TR, Fitzgerald DJ, Moss B, Pastan I, Berger EA (1988) Nature 335:369-372
12. Till MA, Ghetie V, Gregory T, Patzer EJ, Porter JP, Uhr JW, Capon DJ, Vitetta ES (1988) Science 242:1166-1168
13. Capon DJ, Chamow SM, Mordenti J, Marsters SA, Gregory T, Mitsuya H, Byrn RA, Lucas C, Wurm FM, Groopman JE, Broder S, Smith DH (1989) Nature 337:525-531
14. Traunecker A, Schneider J, Kiefer H, Karjalainen K (1989) Nature 339:68-70
15. Stüber D, Matile H, Garotta G (1990) In: Immunological Methods, Vol IV, in press

16. Certa U, Bannwarth W, Stüber D, Gentz R, Lanzer M, Le Grice S, Guillot F, Wendler I, Hunsmann G, Bujard H, Mous J (1986) EMBO J 5:3051-3056

17. Hochuli E, Döbeli H, Schacher A (1987) J Chromatography 411:177-184

18. Sauermann U, Schneider J, Mous J, Brunckhorst U, Schedel I, Jentsch KD, Hunsmann G (1990) AIDS Res human Retrov 6:813-823

19. Summers MD, Smith GE (1987) A manual of methods for baculovirus vectors and insect culture procedures. Texas AES Bulletin n° 1555, Publishers College Station, Texas

20. Dirckx L, Lindemann D, Ette R, Manzoni C, Moritz D, Mous J (1990) Virus Research, in press

21. Moritz D, Dirckx L, Mous J, Schneider J (1990) FEBS Lett, in press

22. Gallati H, Pracht I, Schmidt J, Häring P, Garotta G (1987) J Biol Regul Homeost Agents 1:109-118

23. Miyoshi I, Kubonishi I, Yoshimoto S, Akagi T, Ohtsuki Y, Shiraishi Y, Nagata K, Hinuma Y (1981) Nature (London) 296:770-771

24. Sattentau QJ, Dalgleish AG, Weiss RA, Beverley PCL (1986) Science 234:1120-1127

25. Mizukami T, Fuerst TR, Berger EA, Moss B (1988) Proc Natl Acad Sci USA 85:9273-9277

Discussion -THE USE OF RECOMBINANT PROTEINS IN AIDS RESEARCH: DEVELOPMENT OF A CD4/GP120 BINDING ASSAY

H.J. Röthig

What about oral availability of the Roche protease inhibitor?

J. Mous

There is some oral bioavailability observed in rats and marmosets. It is not yet known whether the compound is orally bioavailable in man.

A. Ganser

What kind of toxicity can be expected or anticipated for the protease inhibitor?

J. Mous

The HIV protease belongs to the class of aspartic proteinases as also the human enzymes pepsin, renin and cathepsin. However, our protease inhibitor displays a very high specificity and has practically no inhibitory effect on the human enzymes. Furthermore the inhibitor shows a very large therapeutic index, because of its high antiviral potency and low cytotoxicity. Preliminar experiments in animals do not show any acute toxicities from treatment with the compound at relatively high doses. However, we have to await the results of the ongoing toxicity studies to give a more precise answer.

R.G. Werner

Have you done a stoichiometric calculation of how many CD4 molecules you would need to inactivate one virus and how many grams of CD4 constructs you would need for one dose, and how much material you would need for the entire treatment of the life cycle of the patient?

J. Mous

There is one problem in that calculations done in the laboratory are mostly done with laboratory strains of HIV. So the calculations based on these viruses are probably not reflecting the real situation in patients. In fact, to prevent clinical isolates to infect T-cells in vitro one needs up to a thousand fold more soluble CD4 than to prevent the infection by a laboratory strain HIV. And the first results of treatment of patients with soluble CD4 seem to reflect that situation because these trials do not show any effect of treatment on any of the parameters of viremia in the treated patients. There is no drop in p24 and there is no increase of CD4 cells. This is not due to antibody neutralization of injected soluble CD4. It seems that in spite of the high doses used in these trials, the amount of protein is still not high enough to efficiently prevent the virus to infect the cell. These viruses are quite different from the ones used in the laboratory.

© 1991 Elsevier Science Publishers B.V. (Biomedical Division)
The clinical pharmacology of biotechnology products.
M.M. Reidenberg, editor

FGF RECEPTORS AS TARGETS FOR DRUG DEVELOPMENT

CRAIG A. DIONNE, MICHAEL JAYE AND JOSEPH SCHLESSINGER

Molecular Biology Division, Rhône-Poulenc Rorer Central Research, King
of Prussia, PA 19406 (USA) and Department of Pharmacology, New York
University Medical Center, New York, NY 10016 (USA)

BACKGROUND AND RESULTS

Clinical Applications for the FGFs

The fibroblast growth factor (FGF) family is comprised of at least
seven closely related small proteins (15-29 kDa) which stimulate growth
of a wide variety of cells of mesenchymal, epithelial, and neuroecto-
dermal origin (1). The first FGFs to be characterized in any detail and
cloned are acidic FGF (aFGF) (2) and basic FGF (bFGF) (3) which were
purified by several different assays including promotion of cell growth
and induction of angiogenesis (1). The newest member of the FGF family,
keratinocyte growth factor (KGF), was purified and eventually cloned on
the basis of its ability to selectively promote the growth of epithelial
cells (4).

In contrast, the remaining four FGFs were identified on the basis of
their ability to transform normal cells: hst/KFGF was isolated as a
transforming gene from human stomach tumor and Kaposi's sarcoma DNA
(5,6); the int-2 gene is activated by the nearby insertion of mouse
mammary tumor virus (7); FGF-5 was isolated as a transforming gene from
bladder carcinoma (8) and FGF-6 was isolated by extensive homology to
hst/KFGF (9). It is worth noting that aFGF and bFGF are also able to
act as transforming factors when overexpressed in appropriate cell
lines (10,11). It is possible that the transforming activity of the
FGFs can contribute to human cancer, since the hst/int-2 locus is
amplified in human breast and esophageal carcinomas (12-15).

The FGFs are important in normal embryonic development since they
induce the early stages of mesoderm formation (16-18) and are expressed
in temporal and spatial specific patterns during embryogenesis (1). In
addition, the FGFs serve as neurotrophic agents and survival factors for
neuronal cells (19,20). All of the FGFs exhibit high binding affinity
towards heparin which, in addition to playing an important role in the
animal, allows easy purification of the growth factors.

Since the FGFs promote angiogenesis and growth, survival and migration of most cell types, they have been proposed as general wound healing agents and for the specific conditions of corneal ulcers and bone fracture repair. Although these represent therapeutic areas for FGFs as agonists, there are many other areas where FGF antagonists might be particularly useful. These include conditions of inappropriate angiogenesis such as diabetic retinopathy and solid tumor vascularization, and conditions of FGF overexpression in primary human tumors (12-15) and benign prostatic hypertrophy (21). Rheumatoid arthritis, which is a complex pathology, exhibits elevated aFGF levels in affected tissues and may respond favorably to FGF antagonists (22). FGF antagonists may also be useful in attenuating herpes virus infection since it has recently been shown that HSV-1 uses the FGF receptor, flg, as a portal of entry into susceptible cells (23). This list highlights only a few of the potential therapeutic applications of FGFs and FGF antagonists and many more can be imagined (24).

In order to facilitate the development of FGF antagonists, we have chosen to clone and overexpress two distinct FGF receptors, flg and bek (25,26). This was necessary because, although most cell types express FGF receptors, their specific identification as particular gene products was unknown. In addition, their low expression (< 5000 receptors per cell) coupled with the high affinity of the FGFs for heparin sulfate proteoglycans found on most cells led to high background in binding assays.

Receptor Tyrosine Kinases

The cloning of full length cDNAs for human flg and bek has been described in detail (27). The FGF receptors are receptor linked tyrosine kinases which have signal transduction characteristics (28) very similar to the EGF and PDGF receptors which activate phospholipase C-γ as part of their signal transduction pathway (29-31). In addition, the earlier studies of EGF and PDGF receptors serve as valuable paradigms for our present characterization of the FGF receptors.

Receptor linked tyrosine kinases can be characterized into four broad families according to their structure (Table I). The first receptor tyrosine kinase (RTK) family, represented by the EGF receptor, consists of an extracellular domain containing two cysteine rich regions, a single transmembrane region and a cytoplasmic tyrosine kinase domain. The insulin receptor family (RTK II) are heterotetrameric structures

consisting of two identical heterodimers. Each heterodimer contains an
extracellular chain containing two cysteine rich regions disulfide
bonded to another chain which has a single transmembrane and a
cytoplasmic tyrosine kinase. The RTK III family, represented by the
PDGF receptor, contains five Ig-like domains in its extracellular
region, along with a single transmembrane sequence, and a cytoplasmic
tyrosine kinase containing an insert of 66-104 amino acids in the middle
of the kinase domain. Each member of each family has an absolute
requirement for tyrosine kinase activity in order to be biologically
functional in signal transduction. In addition, all the RTKs appear to
utilize receptor dimerization as a mechanism of receptor activation.
Preliminary experiments indicate that these rules will hold for the FGF
receptors as well.

The FGF receptors comprise a fourth RTK family which contains 3
Ig-like domains in its extracellular region with an acidic region
located between the first and second Ig-like domains (27,32,33). The
cytoplasmic tyrosine kinase domain is interrupted by a kinase insert
similar to those of the PDGF receptor family. However, the kinase
inserts of the FGF receptor family consist of only 14 amino aids as
opposed to the much larger inserts of the RTK III family.

TABLE I

RECEPTOR TYROSINE KINASE FAMILIES

Family	Receptor Tyrosine Kinases
RTK I	EGF-R, HER2/neu HER3/c-erbB-3, Xmrk
RTK II	Insulin-R, IGF-1-R, IRR
RTK III	PDGF-R-A, PDGF-R-B, CSF-1, c-kit, flt
RTK IV	flg, bek, CEK2

The classification is adapted from Ullrich and Schlessinger
(32) and is extended by the addition of CEK-2 (42) and flt
(43).

The FGF receptors exhibit heterogeneity in their ligand binding
domains. Although flg and bek were initially isolated as forms
containing 3 Ig-like domains (27,33,34), several groups have
characterized flg and bek forms which are missing the first Ig-like

domain, but which are proficient in binding (35-37). Direct comparison between forms containing either 2 or 3 Ig-like domains indicates that, at least for aFGF and bFGF, binding affinity is equivalent between the short and long forms (36). In addition, cDNAs coding for soluble, secreted forms of extracellular domains of flg and bek have been described (36,38) and may be important in regulating the FGF binding or signaling in vivo. A similar secreted form of the EGF receptor extracellular domain has been reported but no specific function for its presence has yet been elucidated (39).

The number of FGFs, together with multiple FGF receptors and various forms of the receptors, indicates that the biological interactions of FGFs and their receptors will probably be quite complex. Nevertheless, it is conceivable that a receptor antagonist which exhibits target specificity can be developed through intelligent drug screening coupled with a knowledge of FGF ligand and receptor biology.

Overexpression and Binding of the FGF Receptors

We generated the FGF receptor overexpressing cell lines, NFlg26 and NBek8, by transfecting NIH 3T3 cells with flg and bek expression vectors, respectively. NFlg26 cells express a predominant flg protein of 150 kDa with minor hyper- and hypo-glycosylated species of 170 kDa and 130 kDa. NBek8 cells synthesize a major bek species of 135 kDa. Treatment of NFlg26 and NBek8 cells with tunicamycin to prevent glycosylation results in a 100 kDa flg product and a 90 kDa bek product which are more consistent with the predicted primary translation sizes of 89 kDa. It has been proposed that glycosylation is necessary for binding of the FGF by the receptor (40). It should be easier to more rigorously address this question with site directed mutagenesis and overexpression now that the receptors have been cloned.

The NFlg26 and NBek8 cells express high levels of specific binding for FGFs as determined by equilibrium binding experiments. NFlg26 and NBek8 cells express a single class of high affinity sites for aFGF, bFGF, and hst/KFGF (27,28). The results of our binding data are summarized in Table II. Both flg and bek exhibit high affinity binding towards aFGF and bFGF (25-80 pM) and bek shows high affinity for hst/KFGF (80 pM). However, flg exhibits less affinity for hst/KFGF (~320 pM) than the other receptor/FGF interactions. Interestingly, a similar 15-fold difference in affinity between bFGF and hst/KFGF for a flg protein lacking the first Ig domain was observed by others (37). We have extended these observations by showing that a bek deletion form, lacking the first Ig-like domain and acidic region, binds aFGF and bFGF with essentially equal affinity as full length bek (36). We can infer from these results that the determinants of binding specificity are contained within the second and third Ig-like domains.

TABLE II

BINDING OF OVEREXPRESSED FGF RECEPTORS

Cell Line	Receptors/Cell	Apparent Kds (pM)		
		aFGF	bFGF	hst/KFGF
NIH 3T3	5,000	60	ND	ND
NNeo4	5,000	60	ND	ND
NBek8	100,000	50	80	80
NFlg26	125,000	25	50	320

Generation of the flg and bek transfected 3T3 cells, NFlg26 and NBek8, and the control neomycin resistant NNeo4 cells is described in reference 27. The apparent dissociation constants were obtained by Scatchard analysis of equilibrium binding data (27,28).

Binding of the three FGFs to either flg or bek results in pronounced activation of the phospholipase C-γ signal transduction system as measured by tyrosine phosphorylation of the receptor and phospholipase C-γ and generation of inositol tri-phosphates (28). These results with cloned, overexpressed receptors, confirm the earlier conclusions that FGF signal transduction proceeds through tyrosine phosphorylation and activation of phospholipase C-γ (41).

CONCLUSIONS

In order to better understand the biology of the FGFs and to develop FGF antagonists, we have cloned and separately overexpressed two full length human FGF receptors, flg and bek. They comprise a separate receptor linked tyrosine kinase family which contains at least one other FGF receptor, CEK2 (42). It is likely that the family will grow as our knowledge of the present members extends our ability to search for other FGF receptors. Two of the receptors bind at least three ligands, which indicates a very high level of redundancy in FGF receptor/ligand interactions. The biological relevance of this redundancy will only be appreciated once the expression patterns and relative affinities of the different components are known. We are presently cloning and expressing other FGFs and FGF receptors in order to complete our picture of relative affinities and are also collecting data on expression in potential therapeutic areas.

The cloning of FGFs and FGF receptors has been critical to our development of FGF antagonists. The overexpressing cell lines allow us to assess binding characteristics of specific identified gene products rather than the potentially mixed FGF receptor populations present on most cells. By removing certain domains, we have been able to determine that the binding activity of bek is contained within the second and third Ig-like domains (36). Further deletions and site specific muta-genesis should help to determine the minimum binding domains on both the ligands and receptors, thus aiding the design of potential antagonists.

From a practical point of view, the much greater signal to noise ratio obtained with these cells lines allows us to perform much more sensitive antagonist screening assays with either adherent cells or solubilized membranes. The FGF receptors in solubilized membranes exhibit similar pharmacology as the FGF receptors in cell monolayers. However, the filter binding assays with solubilized membranes are much more economical in terms of labor and cost of supplies and offer the best choice for primary screens of drug candidates.

Finally, the overexpressing cell lines serve as valuable reagents for the generation and screening of monoclonal antibodies, which may be potentially useful as antagonists in themselves, or at least useful for demonstrating the actual therapeutic utility of small molecule FGF receptor antagonists.

The strategies that we have employed in our work on FGF receptors should be generally applicable to other receptors which are potential drug targets. Naturally, the receptor antagonists which evolve from this program must be tested in appropriate animal models. The pleiotrophic effects of the FGFs, along with the high redundancy in the FGF system, offer very challenging opportunities in drug design.

ACKNOWLEDGEMENTS

The authors would like to thank the Esteve Foundation for the opportunity to present this work and Robin McCormick for her excellent assistance in the preparation of the manuscript.

REFERENCES

1. Burgess WH, Maciag T (1989) Ann. Rev. Biochem. 58:575-606

2. Jaye M, Howk R, Burgess W, Ricca G, Chiu I-M, Ravera MW, O'Brien SJ, Modi WS, Maciag T, Drohan WN (1986) Science 233:541-545

3. Abraham JA, Whang JL, Tumulo A, Mergia A, Friedman J, Gospodarowicz D, Fiddes JC (1986) EMBO J. 5:2523-2528

4. Finch PW, Rubin JS, Miki T, Ron D, Aaronson SA (1989) Science 245:752-755

5. Yoshida T, Miyagawa K, Odagiri H, Sakamoto H, Little PFR, Terada M, Sugimura T (1986) Proc. Natl. Sci. USA 84:7305-7309

6. Delli-Bovi P, Curatola, AM, Kern FG, Greco A, Ittmann M, Basilico C (1987) Cell 50:729-737

7. Dickson C, Peters G (1987) Nature 326:833

8. Zhan X, Bates B, Hu X, Goldfarb M (1988) Mol. Cell. Biol. 8:3487-3495

9. Marics I, Adelaide, J, Raybaud F, Mattei MG, Coulier F, Planche J, de Lapeyriere O, Birnbaum D (1989) Oncogene 4:335-340

10. Jaye M, Lyall RM, Mudd R, Schlessinger J, Sarver N (1988) EMBO J. 7:963-969

11. Rogelj SA, Weinberg RP, Fanning P, Klagsbrun M (1988) Nature 331:173-175

12. Zhou DJ, Casey G, Cline MJ (1988) Oncogene 2:279-282

13. Varley JM, Walker RA, Casey G, Brammar WJ (1988) Oncogene 3:87-91

14. Lidereau R, Callahan R, Dickson C, Peters G, Escot C, Ali IU (1988) Oncogene Res. 2:285-291

15. Tsuda T, Nakatani H, Matsumura T, Yoshida K, Tahara E, Nishihira T, Sakamoto H, Yoshida T, Terada M, Sugimura T (1988) Jpn. J. Cancer Res. (Gann.) 79:584-588

16. Slack JMW, Darlington BG, Heath JK, Godsave SF (1987) Nature 326:197-200

17. Kimelman D, Abraham JA, Haaparanta T, Palisi TM, Kirschner MW (1988) Science 242:1053-1058

18. Paterno GD, Gillespie, LL, Dixon MS, Slack JMW, Heath JK (1989) Development 106:79-83

19. Wagner JA, D'Amore P (1986) J. Cell Biol. 103:1363-1367

20. Anderson KJ, Sam D, Lee S, Cotman CW (1988) Nature 332:360-361.

21. Mydlo JH, Michaeli J. Heston WDW, Fair WR (1988) The Prostate 13:241-247

22. Sano H, Forough R, Maier JAM, Case JP, Jackson A, Engleka K, Maciag T, Wilder RL (1990) J. Cell Biol. 110:1417-1426

23. Kaner RJ, Baird A, Mansukhani A, Basilico C, Summers BD, Florkiewicz RZ, Hajjar DP (1990) Sciene 248:1410-1413

24. Lobb RR (1988) Eur. J. Clin. Invest. 18:321-336

25. Ruta M, Howk R, Ricca G, Drohan W, Zabelshansky M, Laureys G, Barton DE, Francke U, Schlessinger J, Givol D (1988) Oncogene 3:9-15

26. Kornbluth S, Paulson KE, Hanafusa H (1988) Mol. Cell. Biol. 8:5541-5544

27. Dionne CA, Crumley G, Bellot F, Kaplow JM, Searfoss G, Ruta M, Burgess WH, Jaye M, Schlessinger J (1990) EMBO J. 9:2685-2692

28. Bellot F, Kaplow JM, Crumley G, Basilico C, Jaye M, Schlessinger J, Dionne C (1990) submitted.

29. Margolis B, Rhee SG, Felder S, Mervic M, Lyall R, Levitzki A, Ullrich A, Zilberstein A, Schlessinger J (1989) Cell 57:1101-1107

30. Meisenhelder J, Suh PG, Rhee SG, Hunter T (1989) Cell 57:1109-1122

31. Wahl MI, Nishibe S, Suh PG, Rhee SG, Carpenter G (1989) Proc. Natl. Acad. Sci. USA 86:1568-1572

32. Ullrich A, Schlessinger J (1990) Cell 61:203-212

33. Lee PL, Johnson DE, Cousens LS, Fried VA, Williams LT (1989) Science 245:57-60

34. Safran A, Avivi A, Orr-Urtereger A, Neufeld G, Lonai P, Givol D, Yarden Y (1990) Oncogene 5:635-643

35. Reid HH, Wilks AF, Bernard O (1990) Proc. Natl. Acad. Sci. USA 87:1596-1600

36. Crumley G, Kaplow JM, Bellot F, Schlessinger J, Jaye M, Dionne C (1990) Submitted

37. Mansukhani A, Moscatelli D, Talarico, D, Lavytska V, Basilico C (1990) Proc. Natl. Acad. Sci. USA 87:4378-4382

38. Johnson DE, Lee PL, Lu J, Williams LT (1990) Mol. Cell. Biol. 10:4728-4736

39. Petch LA, Harris J, Raymond VW, Blasbland A, Lee DC, Earp HS (1990) Mol. Cell. Biol. 10:2973-2982

40. Feige JJ, Baird A (1988) J. Biol. Chem. 263:14023-14029

41. Burgess WH, Dionne CA, Kaplow J, Mudd R, Friesel R, Zilberstein A, Schlessinger J, Jaye M (1990) Mol. Cell. Biol. 10:4770-4777

42. Pasquale EB (1990) Proc. Natl. Acad. Sci. USA 7:5812-5816

43. Shibuya M, Yamaguchi S, Yamane A, Ikeda T, Tojo A, Matsushime H, Sato M (1990) Oncogene 5: 519-524.

Discussion - FGF RECEPTORS AS TARGETS FOR DRUG DEVELOPMENT

A.J.H. Gearing

Can you review the data that supports the oligomerization model for signal transduction?

C. Dionne

We have data concerning the FGFs and the PDGFs receptors. If we take ^{125}I labeled acidic FGF and cross link it to cells we do not see dimerization, but if we do that same experiment and add DSS as a cross-linking agent we see the dimers. This has been done not only in just over-expressing cell lines but with soluble extracellular secreted forms. So it works independently of the cell membrane as well.

M.M. Reidenberg

Would you comment on where you see the selectivity in this system in the context of future drug development?

C. Dionne

That is why we are looking at the different binding domains of FGF receptors. The biggest problem is that these receptors are everywhere. Potential points of antagonist development are the kinase domain as well, but these are probably the least selective. We only have minor selectivity differences with HST binding to flg. Flg is expressed almost everywhere we look, and so is bek, although at much lower levels, except in embryonic cells where the expression is much higher. We are looking at the CEK2 gene, which is another FGF receptor, and that I think will be much more limited.
So we are at the very early stages.

R.G. Werner

You mentioned that the FGF has a whole range of activities, including angiogenesis and promotion of growth. If these compounds were used for wound healing, how would you see the problem of tumor promotion in treated patients?

C. Dionne

This is a very interesting point. The tumorigenicity only applies to cells that have set up an autocrine type system. One cannot get tumor formation in whole animals by injecting them with FGFs. However, cells that have been transfected and express their own FGFs are chronically stimulated in a different way than chronic addition and tumorigenesis does then occur.

A.J.H. Gearing

Can you give us an idea of the levels of FGFs in say a healing wound or in an RA synovial fluid?

C. Dionne

The estimates are around ng/ml. This is one of the reasons why we are looking at antagonists rather than at agonists, although its exogenous addition does work. However, FGF 5 is everywhere and at relatively higher levels.

© 1991 Elsevier Science Publishers B.V. (Biomedical Division)
The clinical pharmacology of biotechnology products.
M.M. Reidenberg, editor

DEVELOPMENT OF ANTAGONISTS FOR IFNγ AND TNF

MICHAEL STEINMETZ

Central Research Units, F.Hoffmann-La Roche Ltd, CH-4002 Basel, Switzerland

Multicellular organisms use peptidic and non-peptidic molecules to transmit signals between organs (hormones) and cells (cytokines). These molecules exert their functions by binding to polypeptidic receptors located on the target cell surface or within the cell. Many useful drugs act by binding to these receptors and thereby altering their biochemical and biophysical properties. So-called agonists evoke a response similar to that produced by the natural ligand, while antagonists block physiological triggering of the receptor.

MANIPULATION OF THE IMMUNE SYSTEM TO TREAT DISEASE

Our understanding of the immune system that protects higher vertebrates against infectious organisms and transformed cells has grown steadily over the past twenty years. Today the essential cellular components, most if not all of the molecules involved in antigen recognition, and many but certainly not all of the cytokines regulating the cellular interactions in the system have been characterized.

The enormous progress that has been made is mainly due to the application of recombinant DNA technology which enables us to identify, clone and produce even rare polypeptides in large amounts, such that their activities can be studied *in vitro* and *in vivo*. In fact, many of the clinically relevant recombinant proteins, already on the market or presently in clinical trials, are factors which play a role in the activation, growth and differentiation of cells of the immune system. These recombinant drugs are currently being used primarily to enhance immune responses against tumor cells and to replenish cells after chemotherapy.

Another family of factors, currently only incompletely characterized, downregulates immune responses to prevent harmful effects on the host itself. Some of these factors might be important for the establishment of tolerance to self components. Although the aetiology of local and systemic autoimmune diseases like multiple sclerosis and rheumatoid arthritis, respectively, is currently not understood, one can assume that mechanisms which normally downregulate immune responses are not functioning properly.

INTERFERENCE WITH IMMUNE RECOGNITION AND EFFECTOR FUNCTIONS

Several possibilities exist to block immune reactions deliberately. Some of these are summarized in Fig.1. One could try to block antigen uptake or processing, interfere with antigen presentation by class I and class II molecules of the major histocompatibility complex (MHC), block recognition of the antigen/MHC complex by T cell receptors, or inhibit effector reactions like cytokine release, expression and binding of adhesion molecules, or binding of cytokines to their corresponding cell surface receptors.

In the following I will concentrate on our efforts to develop antagonists for interferon γ (IFNγ) and tumor necrosis factor (TNF), potentially useful for the treatment of autoimmune diseases. Primarily papers from our own laboratories will be cited, more complete reference lists can be found in these publications.

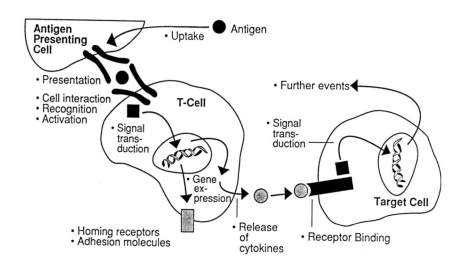

Fig.1 Immune recognition and response reactions provide multiple targets for interfering with autoimmune disease

APPROACHES TO THE IDENTIFICATION OF IFNγ and TNF ANTAGONISTS

We are using both random drug screening and rational drug development programs for the identification of antagonists. All programs involve the use of recombinant forms of the receptors and their ligands. For random screening of natural and synthetic compounds, a binding assay with purified recombinant receptor and ligand is being used (Fig.2). For rational drug development, the molecular interactions in the receptor/ligand complex are being analysed so that small molecules able to block binding can be designed (Fig.3). Other approaches to the identification of receptor or ligand antagonists are also possible, some of which will be mentioned below.

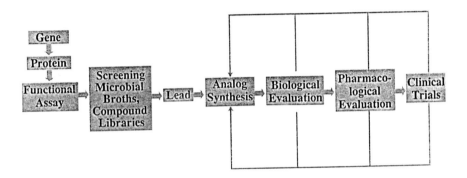

Fig.2 Random drug screening approach for finding low molecular weight antagonists

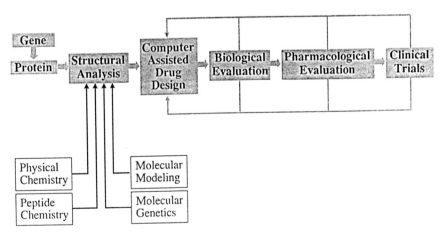

Fig.3 Rational drug development of low molecular weight antagonists

IFNγ ANTAGONISTS MIGHT BE USEFUL FOR THE TREATMENT OF ACUTE AND CHRONIC INFLAMMATORY DISEASES

The rationale for the development of IFNγ and IFNγ receptor antagonists relies on experimental observations made in animal models of various human diseases. In these models it was shown that antibodies neutralizing IFNγ could be used to prevent or treat certain inflammatory diseases or to prolong the rejection of organ transplants. These findings are relevant for certain clinical syndromes as shown in Table 1.

TABLE 1
POTENTIAL USE OF IFNγ ANTAGONISTS IN ACUTE AND CHRONIC INFLAMMATION

Experimental findings in animals with anti-IFNγ antibodies	Relevant clinical syndrome	Reference
Inhibits LPS-induced inflammation	Local gram-negative infections	Heremans et al. (1)
Prevents LPS-induced shock	Septic shock	Heremans et al. (2)
Prevents cerebral malaria	Cerebral malaria	Grau et al. (3)
Decreases lupus-like nephritis and adjuvant arthritis	Autoimmunity	Jacob et al. (4,5)
Delays rejection of tumor, skin and heart allografts	Organ transplantation	Landolfo et al. (6) Didlake et al. (7)

IFNγ EXERTS ITS BIOLOGICAL EFFECTS BY BINDING TO A SINGLE CHAIN CELL SURFACE RECEPTOR

Human IFNγ is a glycosylated protein of 143 amino acids in length, encoded by a single gene on chromosome 12. It has no similarity to IFNα or ß and acts in a species-specific way. IFNγ produced by T cells is the major macrophage activating molecule. Through this and other activities it potentiates immune responses. However, IFNγ has direct antiviral and antiproliferative activities as well.

IFNγ induces a biological response by binding to a single chain cell surface receptor of 90 kD in molecular weight (8). The human IFNγ receptor shows ubiquitous expression with 10^3-10^4 copies per cell surface (9). It is encoded by a single gene located on chromosome 6q and is composed of 472 amino acid residues which are about   equally divided between the extra- and intracellular space (10). The receptor binds human IFNγ in a species-specific way with a dissocation constant of 0.1 nM.

A SOLUBLE FORM OF THE HUMAN IFNγ RECEPTOR IS USED FOR DRUG SCREENING AND THREE-DIMENSIONAL STRUCTURE DETERMINATION

We have characterized the entire extracellular region of the human IFNγ receptor by epitope mapping (11) and expressed it in *E.coli* (12) and insect cells (13). Large amounts of recombinant, soluble receptor can be expressed in *E.coli* and purified to homogeneity. After renaturation the recombinant receptor binds IFNγ with an affinity that is only about 10-fold lower than that of the native, membrane-bound receptor. With the recombinant material from *E.coli* we have developed a solid phase receptor binding assay which is currently being used to screen small synthetic and natural compounds for antagonistic activity (14).

From insect cells, infected with a recombinant baculovirus coding for the extracellular region of the human receptor, milligram amounts of recombinant, soluble receptor can also be isolated. The recombinant receptor from insect cells is glycosylated (although differently as compared to the native receptor) and binds IFNγ with essentially the same affinity as the native receptor. This material will be used for crystallization purposes to elucidate the three-dimensional structure of the IFNγ binding site and to start a drug design program.

AN ANIMAL MODEL TO TEST POTENTIAL HUMAN IFNγ ANTAGONISTS

Because of the species specificity it is unlikely that a potential antagonist for human IFNγ, identified by drug screening or drug design, will antagonize murine IFNγ. To be able to test drug candidates in small animals, we are developing a transgenic mouse model. Transgenic mice have been obtained with the human IFNγ receptor gene and shown to express the human receptor on spleen cells and to a lower extent on thymocytes. The transgenic receptor binds human but not mouse IFNγ as expected. Binding, however, is not sufficient to induce a response in transgenic cells as measured by antiviral activity and induction of class I and class II MHC gene expression.

These findings are in agreement with earlier reports showing that the expression of the human IFNγ receptor on mouse cells, achieved by somatic cell hybridization (15) or gene transfer (10), did not confer responsiveness to human IFNγ although binding was obtained. It seems that at least one additional polypeptide, apparently encoded on human chromosome 21, interacts with the human receptor in a species-specific way and is required for signal transduction (15). Once the cloning of the signal transducer has been achieved we will introduce the human gene into our receptor transgenic mice to develop an animal model that can be used to test potential IFNγ antagonists.

CAN THE SOLUBLE RECEPTOR ITSELF BE USED AS AN IFNγ ANTAGONIST?

To test the feasibility of using the soluble receptor itself as an antagonist, we have cloned the mouse IFNγ receptor (16) and expressed the entire extracellular region of the mouse IFNγ receptor using the baculovirus expression system (13). The recombinant receptor has been purified to homogeneity in milligram amounts and shown to bind mouse IFNγ with the same affinity as the membrane-bound receptor (17). The soluble receptor is able to block the induction of an antiviral response by mouse IFNγ in cultured cells.

We are currently testing *in vivo* stability, immunogenicity and clearance of the soluble receptor after injection into mice. We will then assess its ability to block B and T cell responses, graft-versus-host reactions, graft rejection and the generation of autoimmune diseases in murine models. Once transgenic mice with a functional human IFNγ receptor have been developed as discussed above, we will also be able to test the soluble human receptor in a small animal model for its potential use in the clinic.

BLOCKING TNF ACTIVITY MIGHT BE BENEFICIAL FOR ACUTE AND CHRONIC INFECTIOUS AND INFLAMMATORY DISEASES

TNFα and TNFß are two structurally related cytokines primarily produced by activated monocytes, macrophages and T cells. Both cytokines play a role in host defense reactions by a variety of activities. It has been shown, for instance, that TNFα induces the expression of adhesion and MHC class I molecules, plays an important role in granuloma formation in mycobacterial infections, causes the necrosis of certain mouse tumors and triggers antiviral activity in synergy with interferons. TNFα and ß bind to the same two TNF receptors (see below) and in general appear to induce the same spectrum of activities. There is substantial evidence that the active form of both cytokines is the trimer.

The efficacy of TNF as an anti-cancer agent is currently investigated in clinical trials. TNF antagonists might be useful for a variety of infectious and inflammatory diseases. Studies with cultured cells and animal experiments have indicated that TNF is a mediator enhancing tissue-injury in rheumatoid arthritis, bacterial meningitis, multiple sclerosis, cerebral malaria and graft-versus-host disease. Systemic TNF plays a role in malaria and septic shock. TNF also induces HIV expression in latently infected T cells and thus may play a role in the spread of the virus. Antagonists for TNF or TNF receptors therefore might find many applications in the clinic.

TWO HUMAN TNF RECEPTORS ARE STRUCTURALLY DISTINCT

We have used biochemical, serological and molecular genetic studies to show that human cells express two distinct cell surface receptor molecules which bind TNF. It was postulated from initial crosslinking experiments that a 55 kDa TNF receptor existed on human Hep2 cells, while HL60 cells were found to express both the 55 and a 75 kDa TNF receptor (18). Subsequently, monoclonal antibodies with specificity for either the 55 or the 75 kDa receptor were raised against partially purified receptor preparations and used to confirm and extend these initial findings (19). Finally with the help of these monoclonal antibodies, the two TNF receptors were purified to homogeneity and partial amino acid sequences were obtained (20). Oligonucleotide primers, synthesized according to the amino acid information, were used to amplify fragments of the corresponding receptor genes by polymerase chain reaction and full-length cDNA clones for both receptors were isolated using these fragments as probes (21,22).

Both human TNF receptors are transmembrane proteins containing 426 (55 kDa) and 439 (75 kDa) amino acid residues with extra- and intracellular regions of similar sizes. Sequence comparisons show that the extracellular regions of both receptors are structurally related to each other and to the nerve growth factor receptor and other cell surface antigens. The most characteristic structural unit is a cysteine-rich sequence motif, repeated four times in the extracellular regions of both human TNF receptors. In contrast , the intracellular regions of the two receptors, which are rich in proline and serine residues, are not related to one another nor to any other known protein. It is therefore possible that the two receptors connect to different signalling pathways and are functionally distinct.

Using the specific monoclonal antibodies described above as tools, clear-cut evidence has been obtained for the differential regulation of the two TNF receptors (22-25). Mitogen activation of human peripheral blood lymphocytes, for instance, strongly induces the expression of the 75 kDa, but not of the 55 kDa receptor (22). The expression of the 75 kDa receptor was found to be induced also in activated B cells on time scales which depended on the type of stimulus (24,25). In one study of activated tonsillar B cells a relatively late induction of the 55 kDa receptor was observed (24).

There is compelling evidence that both receptors are functional and transduce different signals. For instance, while B cell proliferation upon IgM treatment for 72 h could be blocked by antibodies directed against the 75 kDa receptor, antibodies directed against the 55 kDa receptor had no effect (24). Clearly, molecular genetic studies are needed to further dissect the functions of the two TNF receptors.

TNF ANTAGONISTS

In experiments similar to those described above for IFNγ antagonists, we are using recombinant forms of the human TNF receptors to develop antagonists. The extracellular region of the 55 kDa TNF receptor has already been expressed in soluble form using eukaryotic expression systems. The recombinant receptor will be used for random drug screening and crystallization. Expression experiments are also underway to produce soluble forms of the 75 kDa TNF receptor. In addition to our attempts to find small molecular weight antagonists, we will test the soluble TNF receptors for their potential clinical use. Given the recent success in the development of random peptide libraries (26), one can also start a search for small peptides able to block binding of TNF to its receptors.

SUMMARY

In this short communication I have summarized our approaches to the development of IFNγ and TNF antagonists for clinical use. Three different avenues are being followed. First, a high-flux receptor-based assay is being used to search for non-peptidic small molecular weight antagonists. Second, structural analysis of receptor/ligand interactions should allow us to embark on rational drug design programs. Finally, soluble forms of the receptors are being tested in animal models of human diseases for their potential clinical use. The observation that soluble forms of IFNγ and TNF receptors are found in the serum and urine of febrile patients suggests that these molecules act as physiological antagonists.

ACKNOWLEDGEMENTS

I would like to thank the members of the IFNγ and TNF receptor groups in our department for their enthusiastic research in the areas described here and Drs. Gianni Garotta, Werner Lesslauer and Richard Pink for comments on the manuscript. Ms Anne Iff is gratefully acknowledged for typing the manuscript.

REFERENCES

1. Heremans H, Dijkmans H, Sobis H, Vandekerckhove F, Billiau (1987) J Immunol 138:4175-4179
2. Heremans H, van Damme J, Dillen C, Dijkmans R, Billiau A (1990) J Exp Med 171:1853-1869
3. Grau GE, Heremans H, Piguet P-F, Pointaire P, Lambert P-H, Billiau A, Vassalli P (1989) Proc Natl Acad Sci USA 86:5572-5574
4. Jacob CO, van der Meide PH, McDevitt HO (1987) J Exp Med 166:798-803
5. Jacob CO, Holoshitz J, van der Meide P, Strober S, McDevitt H (1989) J Immunol 142:1500-1505
6. Landolfo S, Cofano F, Giovarelli M, Prat M, Cavallo G, Forni G (1985) Science 229:176-180
7. Didlake RH, Kim EK, Sheehan K, Schreiber RD, Kahan BD (1988) Transplantation 45: 222-223
8. Fountoulakis M, Kania M, Ozmen L, Lötscher HR, Garotta G, van Loon APGM (1989) J Immunol 143:3266-3276
9. van Loon AGPM, Ozmen L, Fountoulakis M, Kania M, Haiker M, Garotta G (1990) J Leukocyte Biology, in press
10. Aguet M, Dembic Z, Merlin G (1988) Cell 55:273-276

11. Garotta G, Ozmen L, Fountoulakis M, Dembic Z, van Loon AGPM, Stüber D (1990) J Biol Chem 265:6908-6915
12. Fountoulakis M, Juranville J-F, Stüber D, Weibel EK, Garotta G (1990) J Biol Chem 265:13268-13275
13. Gentz R, Hayes A, Damlin K, Fountoulakis M, Lahm HW, Ozmen L, Garotta G (1990) submitted for publication
14. Ozmen L, Fountoulakis M, Stüber D, Garotta G (1990) Biotherapy, in press
15. Langer JA, Pestka S (1988) Immunology Today 9:393-400
16. Hemmi S, Pehini P, Metzler M, Merlin G, Dembic Z, Aguet M (1989) Proc Natl Acad Sci USA 86:9901-9905
17. Fountoulakis M, Juranville JF, Gentz R, Schlaeger EJ, Manneberg M, Lahm HW, Ozmen L, Garotta G (1990) submitted for publication
18. Hohmann H-P, Remy R, Brockhaus M, van Loon APGM (1989) J Biol Chem 264:14927-14934
19. Brockhaus M, Schönfeld, HJ, Schläger EJ, Hunziker W, Lesslauer W, Lötscher H (1989) Proc Natl Acad Sci USA 87:3127-3131
20. Lötscher HR (1990) In: Aggarwal BB, Vilcek J (eds) Tumor necrosis factors: Strucutre, function and mechanism of action. Marcel Dekker, New York, in press
21. Lötscher H, Pan YCE, Lahm HW, Gentz R, Brockhaus M, Tabuchi H, Lesslauer W (1990) Cell 61:351-360
22. Dembic Z, Lötscher H, Gubler U, Pan YCE, Lahm HW, Gentz R, Brockhaus M, Lesslauer W (1990) Cytokine, in press
23. Hohmann HP, Brockhaus M, Baeuerle, P, Remy R, Kolbeck R, van Loon APGM (1990) J Biol Chem, in press
24. Heilig B, Mapara M, Brockhaus M, Möller A, Dörken B (1990) submitted for publication
25. Erikstein BK, Smeland EB, Blomhoff HK, Funderud S, Lesslauer W, Espevik T (1990) submitted for publication
26. Scott JK, Smith GP (1990) Science 249:386-390

Discussion - DEVELOPMENT OF ANTAGONISTS FOR IFN-GAMMA AND TNF

M.M. Reidenberg

You mention using antibodies as tools to see if antagonism would be useful. In many cases of acute illness, where one or two doses of drug might suffice, wouldn't the antibody or the FAB fragment of an antibody potentially be a useful therapeutic agent, avoiding the need to develop a small molecule?

M. Steinmetz

As long as one considers an acute response, I would think an antibody might be useful. However if you talk about chronic inflammatory diseases then of course you will not be able to use the antibody.

P. Tanswell

Could you comment on any particular advantages of this expression system SF9 baculovirus?

M. Steinmetz

In our hands the SF9 system is a good system for rapid expression of a molecule which requires a higher eukaryotic expression system. It takes only a few weeks in order to get reasonable amounts of soluble receptor, as in our example, or any other protein which you might wish to express. If you are interested in very high level expression in the end then probably the CHO system is better, but it takes several months to amplify the introduced gene and develop a stable cell line secreting large amounts of the protein of interest.

L. Gauci

Do you think, from the molecular biologist's point of view, that there is going to be any difference in being able to interact with a receptor at the level of a tiny molecule as opposed to a relatively specific large protein?

M. Steinmetz

It is very difficult to answer this question at this point in time. I would think one should be able to develop a small molecular weight molecule which will block binding of the ligand even if the ligand is a high molecular weight protein.

L. Gauci

But our attitude toward the use of proteins as therapeutic agents has changed in recent years. For instance, four or five years ago many people were convinced that multiple injections of monoclonal antibodies would not be possible. Therefore, it would seem logical to gather evidence of the usefulness of protein antagonists before embarking on complex chemical programs.

© 1991 Elsevier Science Publishers B.V. (Biomedical Division)
The clinical pharmacology of biotechnology products.
M.M. Reidenberg, editor

PROTEINS IN SEARCH OF A DISEASE

Leon Gauci
*Department of Clinical Research, F. Hoffman La-Roche Ltd.,
4002 Basel, Switzerland*

Introduction

It's now not clear to me what on earth I was thinking when I accepted to talk about this subject. However, given that this was requested by the Esteve Foundation in the context of their symposium entitled :"*The clinical pharmacology of biotechnology products*" and that many of you are pharmacologists, I've somewhat nåively assumed that at least part of the subject of the title concerns pharmacology.

Are protein drugs really so pharmacologically different from organic chemical drugs? Perhaps not as much as some would think and even then not in the way implied by "looking sperendipitiously for an indication "!.

Let us explore what we understand by disease specificity and how conventional medicines are used. By comparing the labelling of many proteins with those of conventional drugs it will be possible to show that similar strenghts and weakness exist regarding their pharmacological predictiveness or not.

In order to discern whether protein drugs are really so different pharmacologically, we shall firstly review together the history and present role of proteins in medical practice, then examine certain elements which affect our attitude not only to drug development but also to their therapeutic application.

Unless otherwise stated, the use of the word protein refers to recombinant produced.

History of Protein Drugs

Non-recombinant protein drugs are as old as contemporary medicine itself if one takes Jenner's attempts at vaccination as *prima voce*. In the meantime many more injectable protein based medicinal preparations have been used and these include:- blood transfusion, plasma and plasma-derived products, Factor VIII, organ extracts , tissues isolated hormones, vaccines, preparations for immunisation and many more.

Even today the level of protein drug development is still very significant both recombinant and conventionally derived. As a rough measure of the level of importance, in the USA, in 1989 of the 37 New Drug Approvals, 23 were conventional (62%) and 14 were protein (38%) [1]. The list of protein drugs is shown in table 1.

This is probably similar to the situation in most developed countries and indicates that the number and proportion of protein drugs being approved for registration is a major component.

Table 1   FDA Approved Protein Drugs in 1989 [1]

Drug	Indication
Epogen*	anemia associated with chronic renal failure,
Alferon N	condyloma acuminatum,
Eminase	acute mycardial infarction,
Fluosol	prevent of diminish myocardial ischaemia during PTCA,
ATnativ	hereditary antithrombin-III deficiency,
Cryoprec. AHF	coagulation factor replacement,
Engerix-B*	prevention of hepatitis-B,
PedVax HIB	prevention of Haemophilus influenza type B,
Vivotif Berna	immunization against typhoid,
Oculinum	blepharospasm and strabismus,
HIVAB HIV-1	screening test for anti-HIV-1 antigen,
HIVAG-1	diagnostic/prognostic test for HIV-1 antigen,
UBI/OlympusHIV-1 EIA	test to detect anti-HIV-1 in serum or plasma,
Histatrol	positive skin test control.

* Genetically engineered.

Although not all embracing the spectrum of indications is broad. Recombinant protein drugs are still in the minority compared to more conventional methods for extracting and purifying protein drugs. Presumably this is being radically changed and it will not be long before most protein drug manufacturing needs will be switched to recombinant technology, except of course for complex drugs such as blood.

The use of protein drugs is not restricted to any field of medicine. Some of the more common protein drugs shown in table 2 reflect the broad medical applications (not indicated). There is no major area of medical practice where proteins are not at one time or another used in therapy.

Table 2   Some Commonly Prescribed Protein Drugs

- Insulin
- Calcitonin
- Vasopressin
- Growth Hormone
- ACTH
- Interferon Alfa
- Interferon Beta
- Interferon Gamma
- Interleukin-2
- Erythropoeitin
- OKT3 MAb
- Anti-lymphocyte serum
- Collagen
- Hyperimmune gamma globulins
- Vaccines
- Plasma Products
- Factor VIII
- Blood Transfusion

This in no way is an exhaustive list but makes the point about the commonness of therapeutic use of especially non-recombinant proteins.

Current research activities are enormous to the point where it is difficult to quantify. More than a hundred and fifty different human and non-human recombinant proteins are currently being evaluated in clinical trials in Europe and the USA. To give a brief overview just the pharmacological categories are given below. In some categories there may be dozens of substances (see table 3).

Table 3   Various Protein Drug Categories Under Clinical Investigation.

- Interferons *
- Growth Factors *
- Promotors of Cellular Differentiation
- Monoclonal Anibodies
- Vaccines
- Hormones*
- Anti-inflammatory
- Modulators of Tissue Matrix
- Immunosuppressives
- Immunopotentiators
- Coagulation Factors*
- Anti-Oncogenes

*Antagonists also being evaluated

Some of the most exciting modern protein drugs are growth factors which include the haematologic growth factors:- erythropoetin, G-CSF, GM-CSF, IL-3, thrombopoetin,IL-2 and IL-1. Among the non-haematopoeitic growth factors worthy of special attention are the epidermal growth factors and their inhibitors, nerve cell and bone growth factors which are likely to lead to important therapeutic agents.

Protein drugs are not new to medicine. They have always been used in the treatment of disease even though the quality of their production left, at times, much to be desired. The advent of recombinant technology has not only considerably increased the scope of discovery and potential application, but has also rendered the whole approach more pharmacologically acceptable. Many modern protein drugs could not emerge from the obscurities of experimental science without the advent genetic engineering.

Mono- and Polymodal Origins of Disease

Many clinicians feel more comfortable treating a disease with a single known aetiology with a specific therapy. Often this is not always possible. Polymodal disease both interms of aetiologies, pathogeneses and multiple non-specific therapies are not rare.

There is a widely held belief that if a disease has a specific cause, more effective therapy is possible compared to developing therapies for polymodal or multi-faceted pathologies. In fact this is not always as simple as it sounds. When the cause of a disease is not known much confusion and uncertainty may dominate the clinical picture including its treatment. When the cause is finally identified and the appropriate specific therapy discovered, all becomes so simple !

To show how such simple considerations affects ones feelings as a clinician an unusual lisiting according to monomodal and polymodal origins of disease is given in table 4.

Table 4   Some examples of Mono- and Polymodal Diseases

Monomodal	Polymodal
Pneumococcal pneumonia	AIDS
Phenylketonuria	Osteoporosis
Pernicious anaemia	Crohn's Disease
Acromegaly	Dementia
Trisomy 21	Diabetes
Scurvy	Rheumatoid arthritis
Xeroderma pigmentosa	Leprosy

Before the spirochaete responsible for syphillis was identified there were some very varied and bizarre explanations for the multitude of clinical syndromes associated with chronic infection. Even after it had been identified but penicillin was not yet discovered, therapies were both complex and troublesome [2].

The lack of an effective treatment for AIDS, even knowing that the disease is caused by HIV virus, is probably clinically more frustrating than treating diabetes with insulin which is effective in controlling some acute aspects of the disease without curing it.

Protein and non-protein drugs are both used to treat monomodal and polymodal diseases.Monomodal diseases treated with proteins, include human growth hormone to treat pituitary dwarfism, erythropoetin to treat EPO deficiency in renal dialysis patients and t-PA to reverse the process of early thrombsis in acute myocardial infarction. Whereas organic chemical drugs are used to treat polymodal diseases such indomethacin in rheumatoid arthritis, tetracycline in acne, steroids in atopic eczema just to mention a few. Similarly alpha interferon is used to treat chronic hepatitis B, chronic myeloid leukaemia and chronic granulomatous disease (polymodal disease and therapy). Anexate reverses benzodiazepine induced sleep [3], penecillin cures pneumococcal pneumonia, antihistamines cures anaphylaxis, all examples of monomodal diseases and organic chemical drugs.

Hence no apparent inherent difference with regard to the simplicity or complexity of the diseases treated by non-protein and protein drugs.

Mechanistic and Specific Target Approach to Drug Development

Some types of drugs emanate from a discovery process that might be discribed as mechanistic whereas others clearly do not (see table 5). Certain pathologic mechanisms are less frequently or not at all used as the primary driving force for drug discovery, often because they are too complex or their applicability unproven. Hence it will be assumed that the discovery of a new anti-inflammatory drug will be active in many different disease states where inflammation is a dominant factor, whereas a drug found to be active against breast tumours may not necessarily be active against lung cancer or leukaemia.

Table 5   Pathologic Mechanisms Used and Unused in Drug Discovery

Commonly Used	Uncommonly Used
Pain	Malignancy
Inflammation	Congenital abnormality
Edema	Ageing
Anaphylaxis	Psychosis
Immunological Adjuvant	Pappillomatosis
Microorganism growth	

Specific target directed drug discovery programs are highly successful if the target is appropriate. Targets can be of all levels of molecular complexity from a heavy metal such as magnesium or calcium, or an enzyme, to a nuclear receptor or a plasma membrane receptor and even an entire organism such as a virus or bacterium. Some examples of these are given in table 6.

Table 6   Examples of Specific Targets for Drug Development.

Organic Chemistry	Biotechnology
Chemotherapy	Gamma interferon receptor *
Specific Proteinase inhibitors	Thrombin
Angiotension converting enzyme	Hematopoietic growth factors
Monoamine oxidase	HIV/CD$_4$ receptor
Prostaglandin synthetase	MAb against cell or soluble antigens
Analogues of active drugs	Hormone & coagulation deficiences
Histaminic receptors *	Epidermal growth factor inhibitors
	Vaccine

* antagonists

Rapid "chemical" progress in both conventional drug and protein based discovery can be derived from this highly specific targetted orientation. Pharmacologic and therapeutic progress depends on the appropriateness of the target to the disease process and whether its inhibition or stimulation is clinically acceptable.

These approaches are used in the discovery of both protein and conventional drugs. The real difference arises with proteins that have species restricted activites, whereby certain animal models or receptor binding experiments cannot be used outside a restricted range of animal species, as is seen with say gamma interferon [4] but not with erythropoetin [5].

There are of course a number of specific fundamental differences on how one discovers different types of proteins, or steroids, or sugars, or protein kinase inhibitors from a chemical viewpoint, but this exceeds the domain of pharmacology.

Modes of Treatment

Surgical practice and thinking had a major impact on early therapeutics. At one time surgical practice was dominated by removal. Removal of diseased or traumatised tissue. Of course this has changed, surgery is now much more sophisticated. Surgeons still spend a lot of their time removing tissues and organs, but in addition they repair, they restore function and they even replace diseased organs or parts of them.

Physicians however did not like to think of themselves as "removers " but rather as modifiers and replacers. Continuing in this very simplistic approach, modern medical treatment may be classified as therapy by:
- Replacement
- Removal
- Blocking, and
- Stimulation,

some examples are given in tabels 7 and 8.

Table 7   **Functional Classification of Medical Therapy: Replacement and Removal.**

Replacement	Removal
Hypothyroidism	Calories in obesity
Pernious anemia	Sodium in essential hypertension
Scurvy	Glutamine in hereditary intolerance
Kwashikor	Tryamine in Migraine
BT in haemolysis	UV light in Xeroderma pigmentosa
Pituitary dwarfism	*Immune Complexes by plasmapharesis*
G-CSF in Neutropenia	*Antigens in allergy*

Footnote: therapies in italics indicate protein therapy.

It would be difficult to say from the examples chosen above that there were major differences between the way protein and non-protein drugs are used, both the pharmacologic and the therapeutic modalities seem to be very similar.

Table 8 **Functional Classification of Medical Therapy: Stimulation and Blocking.**

Stimulation	Blocking
Caffeine in Narcolepsy	ACE inhibitors in Hypertension
Ephedrine in Asthma	H_2-receptor in gastro-duodenal ulcers
Isoprenaline in Asthma	H_1-receptor in allergy
Bisacodylum in constipation	Proton pump blocker in GI ulcers
Androgens in Wasting diseases	5HT blockers in Migrain
GTN in angina pectoris	Pilocarpine in Glaucoma
LHRH agonist in Ca Prostate	ß-blockers in hypertension and anxiety
Interferon alfa 2 antiviral	Heparin in blood coagulation
	IL-1 or G-IFN receptor antagonists

Footnote: therapies in italics indicate protein therapy.

Similarly if one compares the package insert labelling of a selection of conventional and a selection of protein drugs (table 9) it really isn't very difficult to find examples of drugs that have a narrow indication label and those that have a broad one. From this viewpoint also it is not possible to describe protein drugs as fundamentally so different in their pharmacologic development.

Conclusions

Protein drugs are no more or no less in search of a disease than their organic counterparts. The availability of meaningful and predictable preclinical and clinical pharmacologic models is equally sparse for both. Modalistic and mechanistic analyses show that protein drugs are in fact evaluated and developed in a similar fashion to traditional pharmaceuticals.

Protein drugs are not new to medicine and in fact have been around for more than a century whereas biotechnologically produced drugs have only been around for just over a decade. They have a broad application in medicine and rather than be dismayed by the advent of genetic engineering, the improved quality and quantity of these protein drugs both as substances as well as preparations, is a real pharmaceutical advantage. Certain types of protein drugs could only exist because of this technological advance. Of course there are new problems, but these are not proving to be insuperable.

Protein pharmacology however, is different in a number of important respects and are worthy of consideration. Protein drugs may modulate specific cellular functions, for example: 1) growth factors not only of the reticulo-endothelial system but probably for every tissue in the body including nervous tissue, 2) anti-oncogenes, 3) suppression of Philadelphia positive clones in CML; whereas others modulate biological

78

Table 9   Types of Package Insert Labelling Real & Imaginary for Conventional and Protein drugs

Drug	Label
ACE Inhibitor	Hypertension and CHF
Levodopa	Parkinson's Disease
Aminophylline	Asthma
G-CSF *EPO*	*Congential & iatrogenic neutropenia* *EPO deficient anaemia*
anti-LPS MAb *anti-TNF MAb*	*Endotoxic shock* *Shock & Cachexia*
5FU	Ca.breast,colo-rectal ca., Pancreatic ca.,carcinomas of H & N., Ca.bladder and ca. Prostate
Prostaglandin Synthetase Inhibitor	Rheumatoid arthritis,Osteo arthritis, Ankylosing spondylitis,acute Gout, Uveitis,Pleurisry,Pericarditis, Bartter's Syndrome, Patent ductus arteriosus in neonates.
Corticosteroids	Anti-inflammatory, Leukaemai and lymphoma, Asthma and Eczema, Urticaria, Granuloma, Hypercalcaemia...
Interferon alfa-2	*Hairy cell leukaemia,* *Chronic myeloid leukaemia,* *Myeloma and lymphoma,Renal Cell Ca.* *Malignant Melanoma,Kaposi's Sarcoma in* *AIDS, Laryngeal pappilomatosis,Chronic* *hepatitis-B,Chronic hepatitis-C,* *Condyloma acuminata.*
Interferon Gamma	*Congenital Granulaomatosis,Rheumatoid* *arthritis, Renal Cell ca., Cutaneous* *Leishmaniasis,Leprosy, AIDS,adjuvant* *carcinoma.*

Footnote: therapies in italics indicate protein drugs.

mechanisms that can also be altered or influenced by organic or even inorganic substances, such as: 1) promotion of cellular differentiation, 2) inhibition of inflammation, and 3) inhibiton of immunity. It is interesting to note that hypothalamic and pituitary "controlling hormones" are peptides whereas the local tissue hormones they control are not. Peripheral "independent hormones" are peptides.

Exogenously administered proteins similar to their endogenous counterparts renders pharmacokinetics and metabolism evaluation particularly challenging in those circumstances where endogenous levels vary physiologically or in disease. At this time protein drugs are administered topically or by injection. This is changing.

An appreciation of the complex cell surface receptor ligand interactions, post-receptor activation events and gene modulating activities of protein drugs is basic to understanding how they work.

Proteins play special roles in nature and disease. Protein drugs are special in many, but not all ways.

Acknowledgement

I am especially grateful to Sergio Enrill and the Esteve Foundation for providing me with a set of circumstances which forced me to reflect on this subject. As usual my colleagues provided me with a wealth of information and lively critique.

References

1.   Mossinghoff G J (1990) New Drug Approvals in 1989, America's Pharmaceutical Research Companies, Pharmaceutical Manufacturers Association.

2.   Home Handbook of Hygiene and Medicine (1880) by J.H.Kellog' MD, 2nd Ed. pp 1297-1298,1584

3   Amrein R, Hetzel W, Hartmann D, and Lorscheid T (1988) Clinical Pharmacology of Flumazenil Eur.J. Anaesthesiology, Suppl.2 1988 65-80

4   Grossberg S E (1987) Interferons : An overview of their biological and Biochemical Properties.Volume 1 CRC Press Inc., Mechanisms of Interferon Actions, Ed. Lawrence M Pfeffer, Florida USA.pp 1-32.

5   Matsumoto T, Endoh K, Kamisango K, Akamatsu K-I, Koizumi K, Higuchi M, Imai N, Mitsui H and Kawaguchi T. (1990)Effect of recombinant human erythropoietin on anticancer drug induced anaemia. Brit. J. of Haematology,75, 463-468.

Discussion - PROTEINS IN SEARCH OF A DISEASE

C. Dionne

I would like to comment on the problems at the earlier stages of drug development with, for example, FGF and FGF receptors. When we first started two or three years ago we were dealing with two FGFs and one receptor. Now it is much more complicated. Establishing a research programme that is directed during the early stages now has become much more difficult because there are so many options and mostly not enough time or hands to develop all those options.

L. Gauci

Yes, it is a problem related to the area in which you are working. Most clinicians would agree this area is very important. How it would actually turn out in a few years time we don't know. For example, supposing somebody comes along with a recombinant thrombopoietin, from the beginning we know what we want to do with that drug. We know what it should do and what it shouldn't do. We know how many patients we would want to be treating. It represents an easier risk investment type of decision. But no one could deny that the type of work that you are doing is very, very important, the full implications of which remain unknown. So, someway your work has to be financed to allow it to proceed until it is known what its therapeutic applications will be. But that is no different to the sort of problems of the people in the early days had working on prostaglandins.

W. Aulitzky

LHRH learned us, that one substance can have two totally different effects. A stimulating effect and a blocking effect. But this has not really anything to do with the substance or with the hormone itself, it has something to do with receptor phenomenon. Don't you think that the misunderstanding about the effect of the different interferons could also be due to the fact that we are dealing with hormones and that we do not fully understand the receptor phenomenon involved in those actions?

L. Gauci

Yes, I entirely agree with your comment and I would take it one step further. Interferons stimulated a very interesting study aimed at understanding what happens at membrane receptors. At one time we had the very naive idea that down-regulation was responsible for loss of activity but this would appear not to be true. In order for something to be active it has to have a receptor. The binding to the receptor however is not enough, it's got to activate it and stimulate post-receptor activation events.

W.M. Wardell

I'm not convinced that the pharmacology and the clinical pharmacology are really "new" at least the principle. Isn't the position still in part that we have new drugs, and we are finding clinical facts about those drugs?

L. Gauci

I think there is a new pharmacology and I don't think it is restricted to proteins. For instance, I am very impressed with what is happening in the area of sugar chemistry and I think we are going to have to learn about the pharmacology and applications of sugars.

J.A. Galloway

I think that an exceedingly important term in the equation of future drug development and medical treatment will be the impact of the results of mapping the human genome. It seems likely to me that this could well be the topic of an Esteve Symposium before the end of this century.

PHARMACOKINETICS OF BIOTECHNOLOGY PRODUCTS

© 1991 Elsevier Science Publishers B.V. (Biomedical Division)
The clinical pharmacology of biotechnology products.
M.M. Reidenberg, editor

ABSORPTION OF THERAPEUTIC PEPTIDES

B. Robert Meyer

The Division of Clinical Pharmacology, North Shore University Hospital, 300 Community Drive, Manhasset, New York, 11030, and the Departments of Medicine and Pharmacology, Cornell University Medical College, New York, N.Y.

BACKGROUND

Advances in molecular biology and synthetic chemistry have now made possible the development, production, and purification of a wide number of novel peptide and protein compounds. These new peptide and protein compounds provide new opportunities for drug therapy. They also pose a particular challenge to our ability to develop new systems for drug administration.

While oral administration is the most attractive route for the administration of any drug, oral delivery of this class of compounds has been extremely difficult. Enzymatic hydrolysis both in the gastrointestinal lumen and at the epithelial surface destroys much of an administered dose. Peptides and proteins are also poorly absorbed across the gastrointestinal epithelium. Removal from the portal circulation prior to reaching the systemic circulation is also significant.(1)

In contrast to many traditional drugs, many peptides will have their optimal therapeutic benefit when they are administered with a goal other than steady state kinetics. The endogenous secretion of most peptides and proteins does not follow steady state kinetics and the traditional goal of steady state concentrations of drug may be innappropriate for therapy with peptides. Most endogenous peptides are secreted in frequent small pulses. These pulses may vary considerably in their frequency and amplitude over the course of a 24 hour period. When pharmacologic therapy is designed to replace or mimic the endogenous secretion pattern, optimal pharmacologic effect may only be achieved when a similar pattern of administration is achieved. The administration of a peptide with a pharmacokinetic profile different than the endogenous compound, may produce a paradoxical effect.(2)

NEW DEVELOPMENTS

Optimal exploitation of the therapeutic opportunities offered by peptide drugs will require the development of novel techniques for administration of these agents. These techniques should ideally provide reliable and efficient absorption of peptides with a capacity for both steady infusion of compound as well as intermittent bolus administration. A variety of efforts have been made to respond to this need. The areas of development are outlined in Table I. While promising work has been done in each of the highlighted areas, in each case significant problems remain.

The ideal route for administration of any drug would be the oral route. Ease of administration and compliance make this extremely attractive. The problems of oral bioavailability can be partially alleviated by alterations in peptide structure designed to render them resistant to hydrolysis, or by disguising peptides in more readily absorbable forms (1). Enalapril and cyclosporin are two peptide compounds in which oral absorption has proved practical. Recent research has suggested that insulin in "chylomicron"-like emulsions accompanied by protease inhibitors may make oral administration of this compound feasable.(3) The use of bioadhesive polymers to cause administered drug to "stick" to the gastrointestinal mucosa, or the use of protective agents to allow delivery of the peptide or protein to the colon, where enzymatic degradation is diminished, and mucosal absorption may be achieved, may allow the development of other oral preparations.(4-6) To date these efforts are very preliminary. It is also worth noting that even if successful, the temporal characteristics of the delivery of the drug might be less than ideal.

TABLE I

Potential Techniques for the Administration of

Peptide and Protein Drugs

Route of Administration	Strengths	Weaknesses
Parenteral		
Depot	Predictable	Zero Order
	Long Duration	Irreversible
	Zero Order	
Pumps	Demand ± Infusion	Expensive
	Predictable	Cumbersome
		Needles
Transdermal	Compliance	Variability/Depots
	Variable Rates	Cutaneous Toxicity
		Antigenicity
Nasal	Compliance	Local Toxicity
	Rapid Onset	Variability
Inhalational	Compliance	Unexplored
Rectal	?	Compliance /Variability
		Needs Enhancers
Opthalmic	Compliance	Variability
Oral	Ideal Route	Time Course/Efficiency
		Is it possible ?

All of the alternatives to the oral route have produced some successes. All also have significant limitations. Depot preparations are more predictable and efficient, but suffer substantially from an inability to vary the rate of drug delivery over time. The use of depot preparations has the attraction of single dose administration for prolonged therapy of 30-90 days, but has the major disadvantage of inflexibility in dosing and committment to steady state kinetics. Infusion pumps avoid this problem, but are expensive, require a committed and sophisticated patient, and despite advances in miniaturization, will remain somewhat cumbersome. Nasal administration has been looked at extensively, and both ADH and IHRH analogues are now commercially available for administration by this route. The extension of this technique to other peptides remains in question. Many larger peptides appear to require the use of an enhancer for significant nasal absorption.(7) A recent report of the successful inhalational administration of the IHRH agonist leuprolide to human volunteers raises the interesting possibility of using the pulmonary alveolar bed for peptide administration.(8) This area merits further investigation.

NEW DEVELOPMENTS IN TRANSDERMAL DELIVERY

Our research has investigated the feasibility of the utilization of the transdermal route for peptide drug administration. We have investigated this route because it offers the attraction of ease of application and good patient acceptance, the possibility of modulation of drug effect by varying the rate of drug absorption, and ease of rapid discontinuation of therapy.

The outermost layer of the human skin is the stratum corneum. This portion of the skin is approximately 10 to 40 cells in thickness, and is composed of a rich extracellular matrix of lipids arranged in a highly ordered fashion. The keratinocytes are immersed like bricks in this extracellular lipid mortar work. The stratum corneum is pierced by hair follicles and eccrine glands, which comprise less than one tenth of one percent of the total skin surface area. The structure of the stratum corneum establishes it as a very effective barrier to the absorption of most chemical entities(9).

Traditional transdermal absorption techniques have used as the driving force for absorption, the passive diffusion of drug along a concentration gradient established across the stratum corneum. Successful passive absorption of drug requires a potent molecule which partitions equally between lipid and water, has a wide toxic therapeutic ratio, and a very low molecular weight. Peptides do not generally fit this description. Therefore, successful transdermal administration will require the modification of the traditional techniques for transdermal absorption. We have sought to overcome these limitations by using an electrical current to alter the pattern of cutaneous absorption, and to provide a means to actively manipulate and control the transdermal absorption of peptides and proteins.

The use of electrical fields to enhance the transdermal transport of a compound is generally referred to as "iontophoretic" delivery. The technique of iontophoretic transport was first described over a century ago, and has been reported on intermittently since that time.(10,11) The traditional understanding of that process held that the technique was effective because the electrical field directly induced the movement of the charged solute (drug) across the stratum corneum. The full utility of this technique was not explored at least in part because an analysis of most drugs suggested that they were not highly mobile in electrical fields, and that successful administration of therapeutic quantities across the skin would require the use of levels of electrical energy that would produce severe and unacceptable cutaneous injury.

Gangarosa and colleagues were among the first to suggest that this simple interpretation of iontophoretic transport was inadequate. He noted that a variety of non-ionic solutes were transported across mouse skin during the process of water iontophoresis. This transport of neutral compounds could not be explained by the use of routine iontophoretic transport theory. Gangarosa termed the process "iontohydrokinesis".(12) Subsequent investigations by Barry,(13) Burnette,(14,15) and most recently by Pikal(16-18) have clarified the nature of the process that is present.

In situations where passive diffusion is small, and where iontophoretic transport of solute is small, the induction of water flow across the skin occurs by electroosmotic and by transport number effect. The transdermal transport of water appears to have two major effects upon transdermal absorption. Hydration of skin alters cutaneous permeability. This occurs separate and apart from the direct effect of an electrical field. Thus, electrically mediated transport can function as a technique for cutaneous hydration, and may produce primary changes in the passive permeability characterstics of the skin. In addition, the magnitude of water flow induced by electroosmosis is sufficient to induce the "convective" transport of solutes dissolved within the water.

Burnette demonstrated that significant transdermal flux of the tripeptide TRH occurred in solutions at pH's of 4 and 8, with current values ranging between 0.1 and 0.5 mA/cm^2. Transport was greater at pH 8 (where TRH is without charge) as compared to pH 4, (where a small portion of TRH is in a charged form). However, even at the lower pH, ionic flow appeared to account for less than one tenth of one percent of the observed charge transport.(19) Similar data for other neutral compounds is also now available.(18)

We have attempted to investigate the practical utility of this technique for the transdermal administration of peptides and proteins. To investigate this technique, we needed a self contained and easily applied electrically powered transdermal patch. This patch (POWERPATCH (R), Drug Delivery Systems Inc., New York, N.Y.) has small, flexible batteries as a source of electrical power, an integrated circuit system to control current, two drug reservoirs at the positive and negative electrodes, and a peripheral adhesive to keep it adherent to the skin. Using this iontophoretic patch system, we have investigated the feasibility of iontophoretic administration of therapeutic doses of peptides to humans.

Leuprolide is nine amino acid polypeptide with a molecular weight of 1209. It is neutral or positively charged at most pH values, having a pka above 8. The carboxyl terminus of the molecule has been replaced with an ethyl amide group. The compound is an LHRH agonist, and is extremely safe for acute administration. This makes the use of human volunteers feasible.

Our initial investigation was a randomized double-blind cross over trial in 13 normal male volunteers. These volunteers were studied on 2 days, 7 days apart. On one study day, they received electrically powered patches containing 5 mg of leuprolide added to the positive electrode. The patches were calibrated to deliver a constant direct current of 0.22 mA. This current was distributed over a surface area of 40 cm^2 at the positive electrode, giving a current density at the positive electrode of 0.005 mA/cm.

On the alternate study day, the subjects received an identical appearing patch containing 5 mg of leuprolide, but without a completed electrical current. The patches were left in place for 8 hours, during which time serum LH levels were obtained to monitor for pharmacologic effect. Mean serum LH levels (Figure 1) showed significant drug effect from active, but not from passive patches. Significant differences between the two arms were seen by 90 minutes (using a significance level of 0.01) and were maintained for the duration of the study. ANOVA also showed highly significant differences (p = 0.0084) and a "therapeutic" response (doubling of baseline LH levels) was seen in 12 of 13 subjects studied.[20]

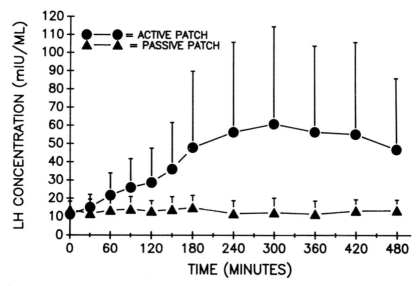

Figure 1: Mean serum LH response (± standard deviation) to transdermal leuprolide administration by "passive" or "active"(iontophoretic) techniques. Reprinted from reference (20) with permission.

In subsequent investigations we have compared the LH response seen with this transdermal administration technique, to the response observed after sub-cutaneous administration. The LH response to sub-cutaneous drug is more rapid peaking as early as 30-60 minutes after the administration of drug. However, the magnitude of LH response as judged by the height of the peak LH response or the AUC for LH response, does not differ between the transdermal and sub-cutaneous routes.(21,22)

Subsequent studies have analyzed serum leuprolide concentrations in individuals receiving patches containing leuprolide at concentrations of 10 mg/ml in either 0.5 or 0.05 molar sodium acetate buffering solution. Again, an electrical current of 0.22 mA and an approximate current density at the positive electrode of 0.005 mA/cm^2 was used. Mean serum leuprolide concentrations were between 400 and 1100 ng/ml for the duration of a 10 hour study. LH response (a doubling of LH levels was seen in 7 of

11 subjects.(23)

TABLE II

Serum Leuprolide Concentrations In Male Volunteers

Current 0.2 mA   Current Density 0.0005 mA/cm^2

Leuprolide 10 mg/ml (0.4 ml)   Na Acetate Buffer 0.5 M or 0.05 M

Responders		Non-Responders
N = 7		N = 4
	Mean LH	
32 mIU/ml	Response	15 mIU/ml
1.18 ± 0.65 ng/ml	Leuprolide Concentration at 150 minutes	0.19 ℁ 0.12 ng/ml
517 ± 179 ng-min/ml	AUC_{0-600}	166 ± 34 ng-min/ml

The patches have been well tolerated by the over 40 individuals we have studied. The level of current used for these studies is well below the cutaneous threshold for pain. Some individuals reported a tingling sensation at the time of patch application. This was always transient in nature. At the time of patch removal small amounts of erythema were noted in approximately one-third of the subjects studied. This erythema resolved in a period of minutes to hours after patch application. The erythema frequently appeared most marked at the periphery of the patch, suggesting that the adhesive rather than the electrical field was the cause of the reaction.

We have also used these types of patches to investigate the iontophoretic transport of insulin.(24) Initial studies in animals investigated the transdermal transport of insulin in albino rabbits with

94

alloxan-induced diabetes mellitus. These animals had patches applied to
their backs for periods up to 24 hours. The animal's skin was prepared
for patch application by close clipping of the hair. Care was taken not
to disrupt the skin and no irritant soaps were applied. Patches
delivering an electrical current of 0.4 mA were used. Serial blood
glucose and serum insulin levels were obtained. A significant rise in
insulin concentration with an accompanying decline in blood glucose levels
was observed. Blood glucose levels had returned to normal within 10
hours. The animals tolerated the patch well. (Figures 2 and 3) No
cutaneous injury was observed.

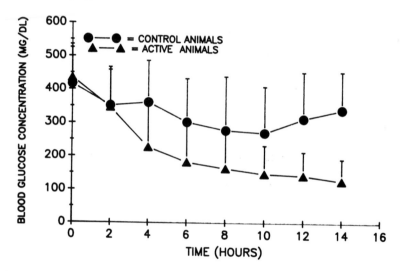

Figure 2: Mean blood glucose levels (± standard deviation) in rabbits
with alloxan-induced diabetes after application of passive or active
(iontophoretic) transdermal patches. (N = 16 active and 8 passive)
Reprinted from reference (24) with permission.

Other investigators have also reported successful "iontophoretic"
transdermal delivery of insulin. Kari utilized an identical animal model,
similar electrical conditions, and achieved substantially higher
concentrations than those that we were able to achieve. However, in this

study, the stratum corneum was partially abraided prior to patch application.(25) This may have substantially altered the permeability characteristics of the skin. Chien and colleagues have also reported a substantial amount of work in a smaller animal model, the hairless rat. Using a system which delivers pulsed direct current, they were able to achieve between 10 and 40 fold changes in calculated permeability coefficients for insulin in the stratum corneum.(26)

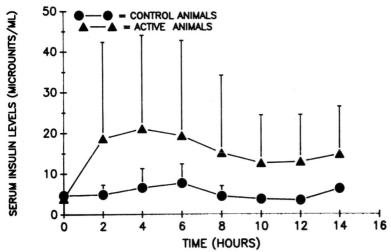

Figure 3: Mean serum insulin levels (± standard deviation) in rabbits from Figure 2. Reprinted from reference (24) with permission.

FUTURE DEVELOPMENT:

The current data support the conclusion that electrically mediated transdermal delivery of peptide drugs is possible. The actual development of a practical system for transdermal peptide delivery still requires considerable investigation. We require a more thorough understanding of the mechanism of transport and the variables in the patch formulation that effect the efficiency and efficacy of delivery. Current understanding suggests that several variables are particularly important.

Current:

The optimal strength of the applied electrical field is not clear. Higher current levels, in most studies, and in theoretical calculations, are associated with improved flow. Chien's data (26) (noted earlier in this discussion) suggest that the use of intermittent pulsed DC current may allow the use of higher current levels with improved cutaneous tolerance.

pH:

Optimal pH conditions for delivery will be dependent upon the specific peptide chosen for delivery. Since most cutaneous pores appear to have a fixed negative charge, the optimal electrode for delivery will generally be the positive electrode, and the optimal pH will have the compound in a neutral or slightly positive state. Once an optimal pH is determined, however, a system for maintaining pH in the appropriate range at each electrode needs to be designed. The use of soluble buffers, such as we have used, will tend to limit the efficacy of drug delivery by providing additional solute molecules which will compete with the drug of interest in the transport process. The ideal buffer may be a system in which buffer is immobilized in a gel matrix at the electrode site.

Concentration:

The concentration of drug at the site of contact with the skin should be maintained at as high a level as practical.

Programmability:

The addition of minor changes in the patch system should allow for the initiation of electrical current (and hence delivery) according to pre-set times, or according to patient activation. Programmable delivery with intermittent bolus therapy is therefore a feasible goal.

LIMITATIONS OF TRANSDERMAL DELIVERY

Transdermal delivery has significant limitations. An examination of the data presented here shows large variability in absorption. No doubt further development and optimization of patch formulations will allow significant reductions in the variability seen in these early studies.

However, significant inter and intra patient variation as compared to traditional parenteral routes of administration may remain. Inter-patient variability is not as critical as intra-patient variability in absorption. Inter-patient variability may be dealt with by the development of several dosage forms, and use of the basic principles for individualization of dose established for oral dosing regimens. A more serious and difficult problem is the issue of intra-patient variability. There is very little information on how individual absorption of drug is effected by repeated applications, environmental conditions such as temperature and humidity, patient activity, and site of patch application.

One of the major issues in electrically-mediated transdermal transport is the occurrence of skin injury secondary to the applied electrical field. The use of current densities in the range reported here is well below the cutaneous pain threshold, and in acute dosing studies no unnacceptable toxicity has been seen. Chronic dosing studies have not been conducted, and will of course need to be performed before any definitive answer concerning the safety of technique can be entertained.

One area of concern with transdermal peptide administration is the potential antigenicity of foreign peptides administered transdermally. Physicians interested in the development of radioimmunoassays have for a long time recognized that the repeated intradermal administration of peptides in appropriate vehicles can lead to the production of antibodies and sensitization of the animal. Whether the transdermal administration of peptides to humans will replicate the experience with peptide and adjuvant administration to rabbits is not known. Whether the development of an antibody response to the administered peptide will effect its therapeutic role is not known, but a cause for concern.

What is the "best" technique for peptide delivery?

The optimal technique for peptide administration will no doubt vary according to the peptide and the therapeutic goals of treatment. In some cases, such as the treatment of prostate carcinoma with an LHRH agent, the use of a once monthly or tri-monthly depot preparation may be a very adequate delivery technique. In other cases, such as the treatment of congenital growth hormone releasing hormone deficiency, optimal results can only be obtained with intermittent nocturnal bolus administration of the drug. We can expect to see a variety of new techniques come to fruition, each with its own particular clinical context and justification.

Acknowledgements:

The research reported here was conducted under a grant from Drug Delivery Systems Inc., New York, N.Y. The animal investigations were conducted at Bio-Dynamics Inc., East Millstone, New Jersey. Co-investigators were James Eschbach, Sanford Rosen, and Dan Sibalis.

References

1. Sternson LA (1987) in Juliano RC (ed) Biological Approaches to the Controlled Delivery of Drugs. New York Academy of Sciences. New York, pp 19-21.

2. Belchetz PE, Plant TM, Nakai Y, Klogh EJ, Krobil E (1978) Science 202:631-633.

3. Cho YW and Flynn M (1989) Lancet 2:1518-1519.

4. Rao S, Ritsche WA (1990) Pharm Res 7:169.

5. Lehr C-M, Bowstra JA, Tukker JJ, et al (1990) Pharmaceutical Research 7:148-168.

6. Brown D, Bae YH, Kim SW (1990) Pharm Res 7:172.

7. Donovan MD, Flynn GL, Amiden GL (1990) Pharmaceutical Research 7: 808-815.

8. Adju A and Garren J (1990) Pharmaceutical Research 7:565-569.

9. Weitz PW and Downing DT (1989) in Hadgraft J; Guy RH (eds) Transdermal Drug Delivery:Developmental Issues and Research Initiatives. Marcel Dekker, Inc., NY pp 1-22.

10. Morton WJ (1898) Cataphoresis or Electrical Medicamental Surgery. American Technical Book Company, NY.

11. Abramson HA, Gorin MH (1945) J. Phys. Chem 44:1094-1102.

12. Gangarosa L, Park N, Wiggins C, Hill J (1980) J. Pharmacol. Exp. Ther. 212:377-381.

13. Barry PH and Hope AB (1969) Biophys. J. 9:700-728.

14. Burnette RR and Bagnufski TM (1988) J. Pharm. Sciences 77:492-497.

15. Burnette RR (1989) in Hadgraft J; Guy RH (eds) Transdermal Drug Delivery: Developmental Issues and Research Initiatives. Marcel Dekker, Inc., NY. 247-291.

16. Pikal MJ (1990) Pharm. Res. 7:118-126.

17. Pikal MJ and Shah J (1990) Pharm. Res. 7:213-221.

18. Pikal MJ and Shah J (1990) Pharm Res. 7:222-229.

19. Burnette RR and Manerro D (1986) J. Pharm. Sci 75:738-742.

20. Meyer BR, Kreis W, O'Mara V, et al (1988) Clin. Pharm. Ther. 44:607-612.

21. Meyer BR, Kreis W, O'Mara V, et al (1989) Clin Pharm Ther 45:129.

22. Meyer BR, Kreis W, Eschbach J, et al (1990) Clin Pharm Ther In Press.

23. Meyer BR, O'Mara V, Eschbach J, et al (1989) Skin Pharm 2:120.

24. Meyer BR, Katzeff H, Eschbach J, et al (1989) Amer. J. Med. Sci. 297:321-325.

25. Kari B, (1986) Diabetes 35:217-221.

26. Chien YW, Siddiqui Y, Sun W, Shi M, Liu JC (1987) in Juliano RC (ed) Biological Approaches to the Controlled Delivery of Drugs. The New York Academy of Sciences, NY, NY pp. 32-51.

Discussion - ABSORPTION OF PEPTIDES

B.P. du Souich

Could you explain the variability in the amount absorbed on the basis of different degrees of sweating?

B.R. Meyer

I don't know. I think that the technique basically is a technique to electrically induce the transport of water and to hydrate the skin. It may be that the amount of hydration of the skin that one achieves will vary with individuals. It may be that the pre-treatment hydration status of the individual will make a difference. I don't fully know what accounts for the inter-individual variation that we saw. I think we need to do further work to clarify exactly what contributes to that.

B.P. du Souich

In my view, a good candidate for this type of administration would be a peptide such as ANP for which one aims at achieving steady state levels, in the treatment of hypertension. Could you comment on that?

B.R. Meyer

In the long run about the first clinical use I would see of this kind of a system, which clearly needs further investigation and development, I would think of an indication where I was not worried about pulsatile delivery. I would want something where I had a steady infusion. I think that is technologically easier to achieve. And then after some sophistication is achieved, turning current on and off and modulating for pulsatile delivery would be a second step in that kind of a development.

J.A. Galloway

What was the charge on the insulin in your studies?

B.R. Meyer

The charge was negative. It was slightly negative and therefore instead of adding it to the positive electrode we actually added it to the negative. In this type of studies one has to consider the pH one is going to work at, the charge of the peptide, and then determine in that context what electrode one is going to add it to, positive or negative. It's something that has to be individualized for each particular peptide, I think.

A.J.H. Gearing

Have you ever tried the technique on patients prone to psoriasis or eczema?

B.R. Meyer

No, for these studies we chose people without known skin disease. Obviously, psoriasis would raise an issue. Most studies say that these patients have increased permeability of their skin rather than diminished, which one might have expected with the exfoliative process.

H.R. Röthig

Your data on leuprolide are quite interesting. We did the same experiments with buserelin. Unfortunately, we found out that the bioavailability is 1 0/00 or less. So this is only 1/10 or 1/20 of what we can administer intranasally. Do you have a good explanation? We think that there is probably a surface or area limitation for absorption. Did you ever try more patches?

B.R. Meyer

I have heard about the data with buserelin and transdermal delivery, but I have not seen the data. I don't know what the differences are between the two systems. In our case, the bioavailability is probably in the range of 5 - 10%. I don't have an explanation for the difference.

S. Erill

Does anyone know whether the transfer of water across the skin is regulated to any extent by ADH. ADH release is influenced by many factors, and this could affect the performance of these devices.

P. du Souich

I do not have precise data concerning the skin, but it is quite clear that the role of ADH is not limited to the kidney.

© 1991 Elsevier Science Publishers B.V. (Biomedical Division)
The clinical pharmacology of biotechnology products.
M.M. Reidenberg, editor

Pharmacokinetics of human tissue-type plasminogen activator

P. Tanswell[1], E. Seifried[2] and J. Krause[3]

Departments of Pharmacokinetics[1] and Biochemistry[3], Dr. Karl Thomae GmbH, D-7950 Biberach;
Med. Klinik und Poliklinik der Universität Ulm[2], D-7900 Ulm (Federal Republic of Germany)

1. INTRODUCTION

Recombinant human tissue type plasminogen activator (t-PA; *Actilyse*[R]) is the most effective agent currently available for coronary thrombolytic therapy of acute myocardial infarction [1]. Physiological t-PA is an endogenous glycoprotein protease that binds to the fibrin component of intravascular thrombi and specifically activates fibrin-bound plasminogen to plasmin, which subsequently dissolves the clot, Fig. 1. Recombinant t-PA administered at pharmacological doses to lyse pathological thrombi causes only limited generation of plasmin in the circulation, avoiding the systemic lytic state and associated coagulation defect which occur during therapy with the first-generation plasminogen activators streptokinase and urokinase.

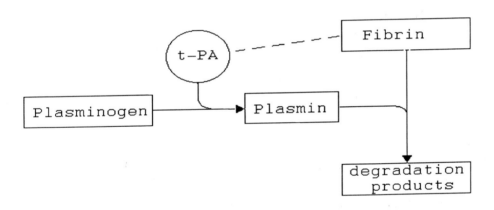

Fig. 1: Mechanism of action of t-PA

The pharmacokinetic properties of t-PA evoked considerable interest early in its development as a thrombolytic agent, when it was discovered that the plasma half-life of the molecule in vivo was only a few minutes due to rapid hepatic elimination [2]. An adequate understanding of t-PA pharmacokinetics is no less important at present, in view of three significant current trends: *(a)* continuing clinical studies aimed at further optimization of dosage regimens [3,21]; *(b)* rapidly advancing knowledge concerning the molecular mechanism of t-PA catabolism [4,22]; and *(c)* substantial ongoing efforts to modify the pharmacological properties of wild type t-PA using site-directed mutagenesis [5 - 7].

2. STRUCTURAL ELEMENTS OF t-PA RELEVANT FOR PHARMACOKINETICS

The t-PA molecule (reviewed in [6]) consists of a single polypeptide chain of 527 amino acids, containing 17 disulfide bridges, Fig. 2. The protease sensitive peptide bond R275-I276 is cleaved by plasmin during fibrinolysis, resulting in a 2-chain form of the molecule in which the amino terminal A-chain remains connected to the carboxy terminal B-chain by a single disulfide bridge. The B-chain comprises a serine protease domain that specifically cleaves the substrate plasminogen.

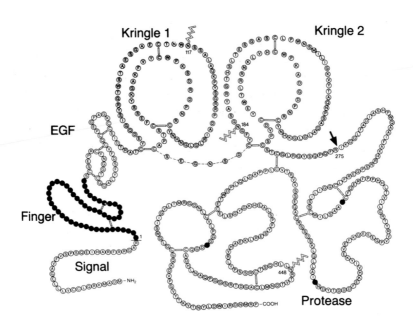

Fig. 2: 2-dimensional structure of human t-PA. N-glycosylation sites are represented by zig-zag lines; the arrow indicates the cleavage site for generation of 2-chain t-PA.

The A-chain mediates other biological functions such as fibrin binding, acceleration of plasminogen activation by fibrin, interaction with the specific inhibitor PAI-1, and binding to cells, including the process of *hepatic clearance*. It contains four autonomous domains which are termed "finger", "EGF" (growth factor), and "kringles 1 and 2", by virtue of their structural homology with other plasma proteins [6]. About 8% of the molecular mass of t-PA ($M_r = 65,000$) is carbohydrate; recombinant t-PA comprises comparable proportions of the carbohydrate variants termed type I and type II. In type I t-PA, asparagine linked carbohydrate side chains are present at N117 (oligomannose) and 184 and 448 (complex oligosaccharides), whereas in type II t-PA N184 is not glycosylated [8].

3. DETERMINATION OF t-PA CONCENTRATIONS DURING *IN VIVO* STUDIES

The form of circulating t-PA in plasma

After intravenous infusion, t-PA appears to circulate in plasma both in its free form and as complexes with the plasma protein inhibitors PAI-1, $\alpha2$-antiplasmin, and C_1-esterase inhibitor [9]. Obviously, the interpretation of pharmacokinetic parameters depends on whether inhibitor-bound or unbound t-PA is detected by the analytical method employed. Further, binding sites for t-PA on platelets and monocytes have been reported. However, t-PA binding to blood cells is probably pharmacokinetically insignificant, since only negligible distribution from plasma into the cellular fraction was found on addition of ^{125}I-t-PA at pharmacological concentrations to whole blood in vitro [10].

Analytical methods

t-PA concentrations in plasma and tissues after administration of the drug in vivo have been analyzed using *(a)* ^{125}I-labeled t-PA and determination of radioactivity precipitated by trichloracetic acid (TCA) or bound by specific antibody, *(b)* functional assays based on plasminogen activation, *(c)* enzyme-linked immunosorbent assays (ELISA), or combinations of these methods.

^{125}I-determination is semi-quantitative and yields meaningful results only up to a few minutes after dosing, since ^{125}I is rapidly cleaved from t-PA by intracellular dehalogenases [11], and TCA and antibodies can also precipitate/bind small peptide fragments. There are however few practicable alternatives for estimating t-PA concentrations in cells and tissues.

Functional assays for t-PA activity in plasma have been numerously reported. The most commonly used are the fibrin plate method, and spectrophotometric assays of plasmin via the chromogenic substrate S-2251 after activation of plasminogen by t-PA in the presence of a noncoagulable fibrin analog such as CNBr fragments of fibrinogen [12,13]. Such assays provide a direct measure of plasmingen activation and are inherently sensitive, but suffer the disadvantage of susceptibility to interference by plasma components. In addition, it is difficult to prevent artifacts caused by loss of t-PA activity via complexation with plasma protein inhibitors *in vitro* in the interval between blood sampling and assay.

2-site ELISA methods measure antigenic mass and utilize a solid phase antibody to capture t-PA from the plasma samples, followed by binding of a second, enzyme labeled antibody for photometric detection via a suitable substrate. ELISA techniques are generally the method of choice due to their high sensitivity, specificity and robustness, and a large number of such assays, utilizing both polyclonal and monoclonal antibodies, has been reported in the literature (e.g. [12]). However, the extent to which a particular ELISA distinguishes between free t-PA and t-PA complexed to plasma protein inhibitors depends on the antigenic determinants recognized by the antibodies used and must be investigated in each case.

4. PHARMACOKINETIC MODEL FOR t-PA: CURRENT CONCEPTS

Route of administration

The high molecular mass of t-PA effectively prevents diffusional transport across biological membranes, hence intravenous bolus and/or infusion is at present the only viable method of delivering t-PA for human therapy. Although intramuscular dosing in combination with absorption enhancers yields thrombolytic plasma concentrations in animal models [14], this route of administration has not yet been investigated in man.

General model

The pharmacokinetic model shown in Fig. 3 can be used to fit all intravenous t-PA plasma concentration/time data so far observed [15], and is equally applicable to t-PA mutants. The model incorporates t-PA bolus or infusion into a central (plasma) compartment postulated to include the

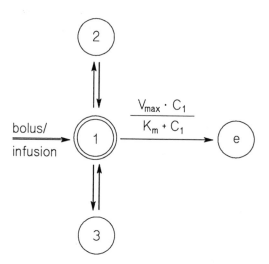

Fig. 3: Pharmacokinetic model for t-PA

liver; first order transfers between the central compartment and 1 or 2 peripheral (tissue) compartments, and capacity-limited (saturable) Michaelis-Menten elimination from the central compartment. The number of compartments and degree of saturation detected in a particular experiment depend on the dosage regimen, the duration of plasma sampling, and the sensitivity of the analytical method.

The elimination rate of t-PA from plasma in this model is given by $-dC_1/dt = V_{max} \cdot C_1/(K_m+C_1)$, where C_1 is the t-PA plasma concentration, V_{max} is the maximum elimination rate and K_m is the Michaelis-Menten constant (value of C_1 for half-maximal elimination rate). Two limiting cases can be distinguished. At very high t-PA plasma concentrations ($C_1 \gg K_m$), elimination is zero order (i. e. independent of C_1), with $-dC_1/dt = V_{max}$. On the other hand, at low plasma concentrations

$(C_1 \ll K_m)$, the more familiar condition of first order t-PA elimination is attained with $-dC_1/dt = k_{1e} \cdot C_1$, where k_{1e} $(=V_{max}/K_m)$ is the elimination rate constant. The effect of nonlinearity (saturation) on the shape of t-PA plasma concentration-time profiles and dose-concentration relationships is illustrated in Fig. 4.

Conditions of nonlinear pharmacokinetics

In a recent study in rats, rabbits and marmosets, using t-PA infusion rates of up to 530 µg/kg/min for 30 min, maximum plasma concentrations (C_{max}) of >110 µg/ml based on ELISA analysis

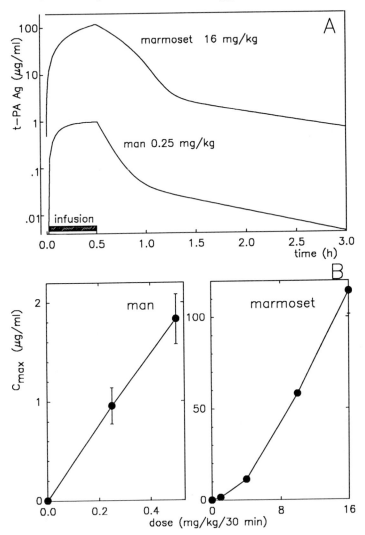

Fig. 4: Effect of nonlinear kinetics (saturation of catabolism) on t-PA plasma concentration profiles (A), and relationship between infusion rate and C_{max} (B). Data based on [13] and [15].

were reached [15]. K_m and V_{max} were computed by simultaneous fitting of multiple plasma concentration-time curves. There was little species variability, with $K_m = 12 - 15$ µg/ml and $V_{max} = 200 - 350$ µg/ml/h. Although a zero order component of t-PA disposition was perceptible in the immediate post-infusion phase at the highest infusion rate in the form of a convexity (Fig. 4A), the nonlinearity became apparent at much lower doses on plotting C_{max} or AUC (area under the curve) against infusion rate (Fig. 4B). Further, in *isolated perfused rat liver*, t-PA elimination capacity was considerably reduced compared to the *in vivo* experiments, with $K_m = 1.5$ µg/ml and $V_{max} = 3.7$ µg/ml/h [15].

The practical conclusion to be drawn from this study is that kinetic data from single doses of t-PA preparations are of limited value. It must be emphasized that the commonly used pharmacokinetic parameters clearance, volumes of distribution and half lives are *not meaningful* under nonlinear conditions, since they are a function of plasma concentration. In particular, reported differences in "clearance" between mutant molecules and wild-type t-PA in animals may be partially attributable to alterations in K_m and V_{max}, resulting in misleading conclusions regarding the potential clinical utility of such variants.

Conditions of linear pharmacokinetics

If the results of the 3 species study can be generalized [15], wild-type t-PA pharmacokinetics in mammals will be effectively linear as long as average plasma concentrations are less than about 10 - 20 % of K_m, i.e. 1.5 - 3 µg/ml. This is the case in clinical studies reported to date. Under these circumstances, the following pharmacokinetic parameters can be calculated from the model (Fig. 3) using standard formulae: volumes of distribution of the central compartment (V_1) and at steady state (V_{ss}), total plasma clearance ($CL = k_{1e}.V_1$), the half lives $t_{1/2\alpha, \beta, \gamma}$ of the α, β- and γ-phases of disposition, and their relative contributions $AUC_{\alpha,\beta,\gamma}$ to the area under the curve.

Evidence for the hepatic elimination of t-PA

The inference of a capacity-limited process for t-PA elimination based on pharmacokinetic data both *in vivo* and in isolated perfused rat liver, with evaluations of K_m and V_{max} [15], confirmed and quantified many independent earlier observations (e.g. [2] and review in [5]), which present strong evidence that the predominant mechanism of t-PA is via hepatic catabolism. In brief, *(a)* increased plasma concentrations of t-PA occur in liver disease, liver transplantation, experimental ligation of the portal vein, liver bypass and partial or total hepatectomy; *(b)* rapid accumulation of t-PA in the liver is observed in tissue distribution studies and whole body autoradiography [16]; *(c)* t-PA has been shown to undergo avid, saturable binding by a hepatic uptake system, followed by intracellular degradation [17].

5. PHARMACOKINETIC PARAMETERS OF t-PA UNDER LINEAR CONDITIONS

Effect of t-PA preparation

This subject has been reviewed [18]. Briefly, the very first pharmacological studies on t-PA were conducted with material derived from Bowes melanoma cell culture. *Recombinant t-PA* (Genentech, Inc.) for initial clinical trials up to early 1985 was produced by a small scale methodology and consisted exclusively of the 2-chain molecule. Subsequently, for pivotal clinical trials and marketing, a large scale manufacturing process was used, yielding t-PA that comprised 70 - 80 % of the single chain molecular form (*Activase*[R]; *Actilyse*[R]). Large scale recombinant t-PA exhibited a higher plasma clearance in myocardial infarction patients than small scale t-PA, which necessitated a dosage adjustment [19]. This alteration in clearance was however clearly shown *not* to be connected with the chainedness of the molecule [18].

Pharmacokinetics in animals

Pharmacokinetics of t-PA have been studied in mice, rats, rabbits, dogs and monkeys [5]. Exact quantitative comparisons between published studies are hindered due to differences in the origin of the t-PA preparations administered and/or in the assay methodologies employed to measure t-PA in plasma. However, the following consistent qualitative picture emerges. After intravenous bolus or infusion of t-PA, plasma concentrations decrease rapidly in an initial, α-phase of clearance characterized by a half-life of < 5 min. The α-phase is dominant, since it encompasses more than two-thirds of AUC. A second, ß-phase of disposition follows with $t_{1/2}$ >10 min, and often a third, γ-phase with $t_{1/2}$ >1 h. This final phase, when observed, represents the elimination of only a few percent of t-PA dose. In a study in which Actilyse[R] was administered to rats, rabbits and marmoset monkeys under identical experimental conditions [15], total plasma clearance of t-PA antigen at low doses was 16 - 23 ml/min/kg, which is close to the hepatic plasma flow rate in these species. V_1 was 46 - 91 ml/kg, corresponding to the plasma volume, and V_{ss} was 230 - 330 ml/kg, indicating moderate tissue distribution and/or binding. $t_{1/2\alpha}$ was 1.1 - 2.4 min with AUC_α= 65 - 77 %, $t_{1/2\beta}$ was 10 - 40 min with AUC_β = 23 - 25 %, and $t_{1/2\gamma}$ was 1 - 1.7 h with AUC_γ = 7 - 10 %.

Pharmacokinetics in man

Clinical pharmacokinetic studies of t-PA satisfying the criteria of adequate blood sample scheduling and computerized data analysis that have been commonplace for more than a decade in investigations of chemical drugs are unfortunately rare. Surprisingly, many earlier studies appeared to use "pencil and graph paper" methods to determine half-lives from post-infusion data although specialized programs such as "TOPFIT" [13] and "NONLIN" [18], which fit preinfusion and multiple infusion plasma concentration data using nonlinear least-squares algorithms, were readily available.

Healthy volunteers. Large scale process t-PA was pharmacokinetically characterized based on plasma antigen and activity in 2 studies at subclinical doses. In the first [12], 8 subjects were administered 0.25 mg/kg t-PA over 30 min; plasma concentrations were compared using fibrin plate, S2251 chromogenic assay, and ELISA methodologies. Clearance based on antigen was 687 ± 63 (SD) ml/min (8.3 ± 0.76 ml/min/kg). The biological assays yielded lower plasma concentrations and thus higher clearances: 1050 ± 104 ml/min (fibrin plate) and 1235 ± 170 ml/min (S2251). However, half lives (2 compartment model) were virtually identical in all assays, with $t_{1/2\alpha}$ = 3.3 - 3.5 min and $t_{1/2\beta}$=26 - 34 min; AUC was 82 - 96 %.

In a further study [13], dose linearity was tested and proved by administering 0.25 and 0.5 mg/kg Actilyse[R] in 2 groups of 6 volunteers (see also Fig. 4). Clearance was equivalent at both doses, the mean value being 620 ± 68 ml/min (8.2 ± 1.3 ml/min/kg) based on t-PA antigen. $t_{1/2\alpha}$ was 4.4 ± 0.3 min, AUC_α 87 ± 1% and $t_{1/2\beta}$ 40 ± 2.7 min. Chromogenic activity analysis yielded CL = 924 ± 160 ml/ min. Comparable results based on ELISA were reported in an earlier study, in which 3 different large scale t-PA preparations were administered to groups of 9 volunteers [18]. In common with the animal studies, CL values in healthy volunteers approximated to hepatic plasma flow [15].

Myocardial infarction patients. In the opinion of the authors, only one study adequately describes pharmacokinetics of the marketed large scale t-PA preparation using the recommen-ded dosage regimen (100 mg over 3 hours; 10 mg as an initial bolus, 50 mg infused in the first hour and 20 mg/h over the subsequent 2 hours) [20]. In 12 myocardial infarction patients, 3-compartment kinetics were observed, referenced both to t-PA antigen and activity. Mean data based on antigen were: $t_{1/2\alpha}$=3.6 ± 0.9 min, $t_{1/2\beta}$= 16 ± 5.4 min, $t_{1/2\gamma}$ = 3.7 ± 1.4 h. AUC_α, AUC_β and AUC_γ were 66, 31 and 3 % respectively, CL was 383 ± 74 ml/min, V_1 2.8 ± 0.9 L and V_{ss} 9.3 ± 5 L. Peak plasma con-centration after the bolus was 3.3 ± 0.95 µg/ml, and steady state concentrations during the ensuing infusions were 2.21 ± 0.47 and 0.93 ± 0.2 µg/ml respectively. A typical data fit is shown in Fig. 5. Similar kinetic parameters were obtained in a novel study using a single 50 mg bolus injection; C_{max} was 9.8 ± 3.6 µg/ml [21].

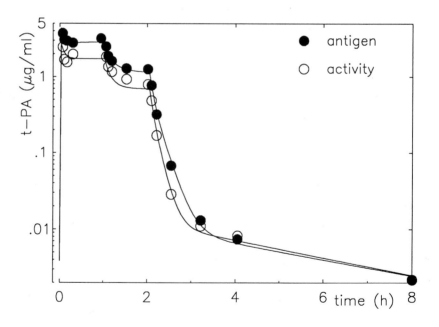

Fig. 5: t-PA plasma concentration-time profile in a myocardial infarction patient (male, 75y); dose regimen given in the text.

6. MECHANISM OF HEPATIC CATABOLISM OF t-PA

Consequent to the pharmacokinetic findings that t-PA is eliminated hepatically, intensive efforts are currently being devoted to elucidate the interaction of t-PA with hepatocytes *in vivo* and *in vitro* using biochemical, cell biological, immunocytochemical and electron microscopy techniques [22, 23]. The present status can be summarized as follows. *(a)* t-PA catabolism is initiated via binding of the molecule to hepatic receptors, followed by internalization in coated pits, degradation by lysosomes, and release of soluble t-PA fragments into the circulation. Urokinase is cleared by a completely different receptor system [4]. Binding of ^{125}I-t-PA to primary rat hepatocytes revealed a high affinity and a low affinity site with K_D values 4 nM and >100 nM respectively, Fig. 6.

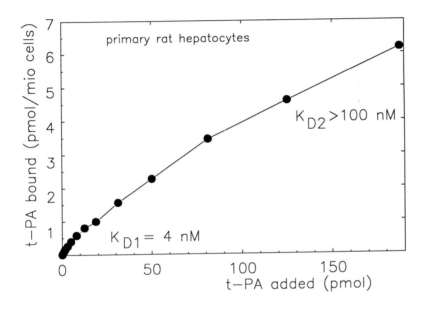

Fig. 6. Saturation binding of ^{125}I-t-PA to primary rat hepatocytes (based on [4]).

(b) Parenchymal, endothelial and Kupffer cell fractions contribute 55 %, 40 % and 6 % to total t-PA uptake in the liver respectively [22]. *(c)* Catabolism in endothelial cells occurs via a mannose receptor recognizing t-PA carbohydrate at N114 (see Fig. 2), and in parenchymal cells via a specific t-PA receptor that recognizes a polypeptide structure localized in the A-chain. Preliminary characterization of the rat hepatic parenchymal receptor indicates that it is located on the cell surface, and is a protein with a molecular mass of about 35000 in SDS gel electrophoresis [24]. The t-PA-specific clearance determinant probably comprises amino acid side chains located in the finger, EGF and kringle 1 domains [25]. The enzymatic active site of t-PA, or other moieties in the B-chain, do not appear to be necessary for recognition by either hepatic receptor system. *(d)* As demonstrated by numerous studies investigating pharmacokinetics of t-PA variants, total plasma

clearance in vivo or in isolated perfused liver is an extremely sensitive indicator of structural modification of t-PA. For example, the single amino acid point mutation R275E, or removal of the high mannose carbohydrate at N114, resulted in a reduction in CL of more than 50 % in both experimental systems [7,26]. Deletion of whole domains reduces the elimination rate of t-PA even more drastically [5]. *(e)* The catabolic fate of t-PA complexes with plasma protein inhibitors, the extent to which these contribute to the overall clearance of t-PA, and the existence (if any) of extra-hepatic clearance pathways for t-PA at pharmacologiocal doses are questions that are as yet unresolved.

7. CONCLUSIONS

The objective of pharmacokinetic studies is to relate doses to pharmacological effects via determination of drug concentrations in plasma and tissues, ultimately enabling prediction of clinical outcome via suitable models. Since the site of action of t-PA is the blood compartment, its therapeutic effect, defined as the rate of thrombolysis, can be expected to correlate directly with t-PA plasma concentrations. Future clinical studies in the setting of acute myocardial infarction should therefore aim at increasing the predictive value of t-PA kinetic parameters, by plasma concentration monitoring and model fit analysis during novel dosage regimens such as front loaded infusions or repeated boluses. In parallel, continuing basic research into the regulation of endogenous t-PA synthesis and catabolism can be expected to yield novel molecular targets for the treatment of chronic thrombotic diseases.

REFERENCES

1. Collen DC, Gold HK (1990) New developments in thrombolytic therapy. Thromb. Res. Suppl. X: 105 - 1312

2. Korninger C, Stassen JM, Collen D (1981) Turnover of human extrinsic (tissue-type) plasmi-nogen activator in rabbits. Thromb. Haemostas. 46: 658 - 661

3. Neuhaus KL, Feuerer W, Jeep-Tebbe S, Niederer W, Vogt A, Tebbe U (1989) Improved thrombolysis with a modified dose regimen of recombinant tissue-type plasminogen activator. J. Am. Coll. Cardiol. 14: 1566 - 1569

4. Krause J, Seydel W, Heinzel G, Tanswell P (1990) Different receptors mediate the hepatic catabolism of tissue-type plasminogen activator and urokinase. Biochem. J. 267: 647 - 652

5. Krause J (1988) Catabolism of tissue-type plasminogen activator (t-PA), its variants, mutants and hybrids. Fibrinolysis 2: 133 - 142

6. Pannekoek H, de Vries C, van Zonnefeld AJ (1988) Mutants of human tissue-type plasmino-gen activator (t-PA): structural aspects and functional properties. Fibrinolysis 2: 123 - 132

7. Krause J, Tanswell P (1989) Properties of molecular variants of tissue-type plasminogen activator. Drug Res. 39: 632 - 637

8. Spellman M, Asa LJ, Leonard CK, Chakel JA, O'Connor JV, Wilson S, van Halbeek H (1989) Carbohydrate structure of human tissue plasminogen activator expressed in chinese hamster ovary cells. J. Biol. Chem. 264: 14100 - 14111

9. Lucore CL, Sobel BE (1988) Interactions of tissue-type plasminogen activator with plasma inhibitors and their pharmacologic implications. Circulation 77: 660 - 669

10. Seifried E, Tanswell P (1987) Comparison of specific antibody, D-Phe-Pro-Arg-CH$_2$Cl and aprotinin for prevention of in vitro effects of recombinant tissue-type plasminogen activator on hemostasis parameters. Thromb. Haemostas. 58: 921 - 926

11. Greischel A, Tanswell P, Busch U, Schumacher K (1988) Pharmacokinetics and biodisposition of recombinant human interferon α-2C in rat and marmoset. Drug Res. 38: 1539 - 1543

12. Seifried E, Tanswell P, Rijken DC, Barrett-Bergshoeff MM, Su CAPF, Kluft C (1988) Pharmacokinetics of antigen and activity of recombinant tissue-type plasminogen activator after infusion in healthy volunteers. Drug Res. 38: 418 - 422

13. Tanswell P, Seifried E, Su CAPF, Feuerer W, Rijken DC (1989) Pharmacokinetics and systemic effects of tissue type plasminogen activator in normal subjects. Clin.Pharmacol.Ther. 46:155-162

14. Sobel BE, Sarnoff SJ, Nachowiak DA (1990) Augmented and sustained plasma concentrations after intramuscular injections of molecular variants and deglycosylated forms of tissue-type plasminogen activators. Circulation 81: 1362 - 1373

15. Tanswell P, Heinzel G, Greischel A, Krause J (1990) Nonlinear pharmacokinetics of tissue-type plasminogen activator in three animal species and isolated perfused rat liver. J. Pharm. Exp. Ther. in press

16. Bakhit C, Lewis D, Busch U, Tanswell P, Mohler M (1988) Biodisposition and catabolism of tissue-type plasminogen activator in rats and rabbits. Fibrinolysis 2: 31 - 36

17. Bakhit C, Lewis D, Billings R, Malfroy B (1987) Cellular catabolism of recombinant tissue-type plasminogen activator. J. Biol. Chem. 262: 8716 - 8720

18. Baughman RA (1987) Pharmacokinetics of tissue plasminogen activator. In: Sobel BE, Collen D, Grossbard EB (eds). Tissue plasminogen activator in thrombolytic therapy. Marcel Dekker, New York, pp 41 - 53

19. Garabedian HD, Gold HK, Leinbach RC, Johns JA, Yasufa T, Kanke M, Collen D (1987) Comparative properties of two clinical preparations of recombinant tissue-type plasminogen activator in patients with acute myocardial infarction. J. Am. Coll. Cardiol. 9: 599 - 607

20. Seifried E, Tanswell P, Ellbrück D, Haerer W, Schmidt A (1989) Pharmacokinetics and haemostatic status during consecutive infusions of recombinant tissue-type plasminogen activator in patients with acute myocardial infarction. Thromb. Haemostas. 61: 497 - 501

21. Tebbe U, Tanswell P, Seifried E, Feuerer W, Scholz KH, Herrmann KS (1989) Single bolus injection of recombinant tissue-type plasminogen activator in acute myocardial infarction. Am. J. Cardiol. 64: 448 - 453

22. Rijken DC, Otter M, Kuiper J, van Berkel TJC (1990) Receptor mediated endocytosis of tissue-type plasminogen activator (t-PA) by liver cells. Thromb. Res. Suppl. X: 63 - 71

23. Seydel W, Stang E, Roos N, Krause J (1990) Endocytosis of tissue-type plasminogen activator by hepatic endothelial cells. Drug. Res., in press

24. Seydel W, Hoffmann H, Krause J (1990) Characterization of the receptor mediating the hepatic catabolism of t-PA. Fibrinolysis 4 (Suppl. 3): A225

25. Bennett WF, Paoni NF, Sethy I, Kerr E, Pena L, Keyt B, Nguyen H, Baldini L, Jones A, Meunier A, Wurm F, Johnson A, Zoller MJ, Higgins D, Botstein D (1990) Mapping the functional determinants of tissue plasminogen activator. Fibrinolysis 4 (Suppl. 3): A37

26. Tanswell P, Schlüter M, Krause J (1989) Pharmacokinetics and isolated liver perfusion of carbohydrate modified recombinant tissue-type plasminogen activator. Fibrinolysis 3: 79 - 84

Discussion - PHARMACOKINETICS OF HUMAN TISSUE-TYPE PLASMINOGEN
ACTIVATOR

H.J. Röthig

We have seen today that by blocking receptors one can influence certain systems. Have you ever thought of blocking the catabolic system in the liver? Because I understand that t-PA is a very extremely expensive drug, you could reduce the amount of drug needed.

P. Tanswell

This is certainly something which could be done. It has been quite clearly shown that there are two pathways for t-PA catabolism. The mannose receptor on endothelial cells is a universal receptor which is responsible for the elimination of plasma glycoproteins in general. It might be quite difficult to block this because of the very large number of receptors available in the body. But what might be more promising would be to block the parenchymal receptor which only gets the t-PA after it has been processed by the endothelial cells, and I think that is probably an approach which could be quite interesting.

H.R. Lijnen

It is known that t-PA has a relatively high affinity for fibrin. Do you think that in a thrombolytic setting adsorption of t-PA to a fibrin clot plays a significant role in its clearance.

P. Tanswell

It would depend on the size of the fibrin. If you regard fibrin as being a compartment then the size of this compartment is probably extremely small. The binding to fibrin both at the site of the clot and circulating and bound to endothelial cells in the vasculature may be responsible for the multi-compartment kinetics which we have seen, in particular the gamma phase. So I agree there is some binding but I don't think it is important from a pharmacokinetic point of view. From the point of view of pharmacodynamics, it could be very important.

J.A. Galloway

Your mentioned that this material is protein-bound. How tenacious is that binding? Is that in the order of IGF 1 or not so strong?

P. Tanswell

It is very difficult to determine this. t-PA is known to bind to alpha 1 antiplasmin. There is also a specific inhibitor (PAI-1) and t-PA also binds to C1 esterase inhibitor. This binding is probably covalent; so it is probably irreversible. However, it is very difficult to determine the extent of binding while avoiding in vitro artifacts, because if one takes a blood sample this binding can continue in vitro. Further, there are conflicting reports in the literature as to what extent t-PA actually circulates in the form of these complexes in vivo. One report states there is very little, whereas another has described quite considerable complex formation.

J.A. Galloway

Are your pharmacokinetics based on total t-PA in the serum?

P. Tanswell

We use a ELISA method which measures t-PA predominantly in the free form. We obtained similar results using the activity assay.

S. Erill

Non enzymatic glycosylation of many proteins occurs in diabetic patients. Do you know whether in diabetic patients the endothelial pathway is, let us say, saturated or behaves just in the same way as in non-diabetic patients?

P. Tanswell

It would depend on whether the glycosylation is of the mannose type. If the glycosylation were of the mannose type then it would be quite possible that there is increased utilization of endothelial cells. If it were of the galactosamine type, these proteins would be eliminated by the parenchymal cells.

© 1991 Elsevier Science Publishers B.V. (Biomedical Division)
The clinical pharmacology of biotechnology products.
M.M. Reidenberg, editor

ROLE OF THE KIDNEY IN ELIMINATING PROTEINS AND PEPTIDES

D. CRAIG BRATER

Clinical Pharmacology Division, Department of Medicine, Indiana University

School of Medicine, 1001 West 10th Street, WD OPW 320, Indianapolis, IN 46202

INTRODUCTION

The kidney plays several roles in the elimination of proteins and peptides. Proteins that are too large to be filtered at the glomerulus must be eliminated by other routes. In contrast, proteins or peptides of molecular size and charge to allow filtration at the glomerulus have appreciable elimination by the kidney; in fact, this elimination may be so rapid as to preclude clinical benefit of the protein. The filtration barrier at the glomerulus restricts proteins of molecular size > 40Å. In addition, the glomerular filtration barrier has a net negative charge. Thus, proteins somewhat smaller than the glomerular pores may still have restricted filtration if they also have a net negative charge. Proteins substantially smaller than the molecular pores are freely filtered even if their charge is negative. The importance of glomerular filtration of recombinant proteins is best illustrated by small molecules that are eliminated so quickly that a therapeutic effect is likely precluded. Examples include α_1-proteinase inhibitor and superoxide dismutase (SOD). Because of the rapid elimination of these two compounds, modifications have been made to the molecule in order to sustain them in plasma for longer periods of time while maintaining pharmacologic activity. Methods for accomplishing this goal will be discussed.

Once proteins and peptides are filtered, catabolism occurs by peptidases at the brush order of the proximal nephron. The resultant small peptides and amino acids are then reabsorbed by the proximal tubule. This reabsorption contributes to isotonic reabsorption of water at the proximal nephron. Two characteristics of this catabolism are important when considering the kidney's role in eliminating proteins and peptides. First, catabolism is saturable as has been demonstrated with SOD. Secondly, this catabolism is inhibitable and so doing allows expression of pharmacologic effects at more distal sites in the nephron. Examples of this phenomenon include atrial naturietic peptide (ANP) and the antibiotic, imipenem.

GLOMERULAR FILTRATION OF PROTEINS AND PEPTIDES

α_1-Proteinase Inhibitor

Native α_1-proteinase inhibitor is primarily eliminated by non-renal routes. However, recombinant α_1-proteinase inhibitor ($r\alpha_1$-PI) derived from yeast is not glycosylated and this form of the enzyme is eliminated by the kidney. In addition, such elimination is so rapid as to preclude therapeutic effectiveness. In order to restrict glomerular filtration, $r\alpha_1$-PI has been conjugated to polyethylene glycol. Polyethylene glycol (PEG) can be synthesized with a variety of sizes and when coupled with a protein can increase the overall molecular size sufficient to impair glomerular filtration and allow a more prolonged presence in plasma. Table 1 shows data demonstrating this phenomenon with $r\alpha_1$PI in mice (1):

TABLE 1: Renal Elimination of Recombinant α_1-Proteinase Inhibitor (from reference #1)

	Half-Life (min)
$r\alpha_1$PI	12
+ Renal Ablation	>60
$r\alpha_1$PI – PEG-2 (M_r = 2000)	110
$r\alpha_1$PI – PEG-4 (M_r = 4000)	>600

The half-life of $r\alpha_1$PI in healthy mice is approximately 12 min. The dramatic increase in half-life with renal ablation indicates that elimination is predominately via renal mechanisms. Coupling $r\alpha_1$PI to PEG having a molecular radius of 2000 (PEG-2) increases the half-life to 110 minutes whereas coupling of the protein to a higher molecular weight form of polyethylene glycol increases the half-life even more (1). Importantly, the pharmacological activity of $r\alpha_1$PI is retained even when coupled to PEG. This strategy of coupling PEG to a protein to increase its molecular size can also be used with other proteins since the conjugation procedure is general (2). Table 2 shows some of the activating reagents that have been used for coupling PEG to proteins and indicates a schematic for the reaction. It is important to note that the coupling reaction may interfere with the pharmacologic activity of the protein itself and that this may differ depending upon the activating reagent.

TABLE 2:   Polyethylene Glycol (PEG) Conjugation with Proteins

---

Couples PEG to the ϵ-amino group of proteins by activating the hydroxyl group on PEG.

1.   Activating reagents

    a.   Cyanuric chloride

    b.   1,1'-carbonyldiimidazole

2.   Synthesis

$$CH_2-(O-CH_2-CH_2)_n-OH$$

|
activating reagent

Activated PEG

|
protein

$$CH_3-(O-CH_2-CH_2)_n-O-\overset{O}{\overset{\|}{C}}-NH-Protein$$

Note: Coupling may impair pharmacologic activity.

Superoxide Dismutase

As indicated above, coupling of PEG to $r\alpha_1PI$ results in retention of pharmacologically active drug in the plasma for a longer period of time. This mechanism has also been used to retain SOD in plasma (2). As shown in Table 3, the half-life of SOD in experimental animals is several minutes, whereas renal ablation extends this half-life considerably. For example, in healthy rats the half-life of SOD is 6 minutes, whereas renal ablation increases this value to 55 minutes, indicating that SOD depends upon the kidney for elimination. This dependence upon renal elimination has also been confirmed by micropuncture techniques and autoradiographic localization of SOD to the proximal tubule (3). SOD can also be coupled to PEG without losing its pharmacologic activity and so doing dramatically increases the elimination half-life as has been demonstrated in mice (2).

TABLE 3: Half-Life of SOD ($M_r = 25\text{Å}$) vs PEG-Modified SOD

	SOD		SOD-PEG
	Normal Renal Function	Renal Ablation	
Rat	6.0 ± 0.5 min	55 ± 5 min	
Mouse	3.5 min		9 hr
			16.5 hr
Man	1.4-3.3 hrs		

Other similar methods have also been used to preclude SOD from glomerular filtration. Molecular engineering techniques have been used to couple two SOD molecules together via a spanning segment (4). So doing increases the molecular size sufficient to prevent renal elimination while retaining pharmacologic activity. Alternatively, SOD has been conjugated to albumin which also accomplishes the same goals (5).

In summary, rapid glomerular filtration of small proteins and peptides may frequently preclude their utility as therapeutic agents. One viable mechanism for restricting glomerular filtration and renal elimination, thereby retaining them in plasma for enough time to exert clinically relevant pharmacologic activity is to derivatize the protein. This can be done in a fashion that will increase the protein's size sufficient to restrict its filtration. Successful employment of this technique has been demonstrated with α_1-proteinase inhibitor and with SOD.

This can be accomplished with polyethylene glycol and other conjugates; it is a promising technique for use with other proteins.

CATABOLISM OF FILTERED PROTEINS

Saturation

Once proteins are filtered by the glomerulus, they are catabolized by peptidases associated with the brush border of the proximal nephron. Amino acids and small peptide fragments are then reabsorbed by the proximal tubule. This catabolic capacity is saturable as has recently been demonstrated with SOD (Brater and Odlind, unpublished data). Studies in animals have shown that elimination of SOD is exclusively by the kidney (3). However, at low doses, only very small amounts, if any, SOD appear in the voided urine. With increasing doses, SOD escapes catabolism at proximal nephron sites and is detected in the urine. This phenomenon creates an interesting anomaly in which total clearance of SOD in man would ordinarily be considered to be via non-renal clearance. This is because the renal clearance is calculated to be negligible since little, if any, drug appears in the urine. This observation is in contrast to the fact that renal elimination is known to be the sole pathway for SOD excretion (3). For example, in healthy volunteers with a dose of 1 mg/kg, the total clearance of SOD is approximately 4L/hr, and based on amounts of SOD appearing in the urine, renal clearance only accounts for 2.6% of total clearance. With increasing doses of SOD and as catabolism in the proximal nephron is saturated, more drug reaches the urine and it thereby appears that a greater proportion of clearance is by renal routes. For example, with a dose of 15 mg/kg, the total clearance of SOD in normal volunteers remains at approximately 4L/hr, but at this dose 52% of clearance can be accounted for by SOD appearing in the urine. In this circumstance, calculated renal clearance is not a valid reflection of the renal contribution to elimination of SOD, but instead represents a mechanism for quantifying the saturability of catabolism of SOD at the proximal nephron.

Inhibition

In addition to being saturable, it is also clear that catabolism of at least some proteins and peptides at the level of the proximal nephron can be

122

inhibited. This has been demonstrated to occur with atrial natriuretic peptide (ANP) (6). Filtered ANP is catabolized by neutral endopeptidase at the brush border of the proximal nephron. As a consequence, ANP's effects at more distal nephron sites are precluded. Recent studies have shown that inhibition of neutral endopeptidase results in pharmacologic effects of ANP more distally (6). As such, increases in urinary volume, sodium, and fractional excretion occur, which correlate with increases in fractional excretion of ANP itself and increases in urinary cyclic GMP, the second messenger for ANP effects:

TABLE 4: Inhibition of ANP Catabolism by the Proximal Nephron* (from reference #6)

	Control	ANP Alone	ANP Plus NEP Inhibition
Volume (ml/min)	0.25	1.18	2.94
Sodium (μEq/min)	52	199	410
FE_{Na} (%)	1.2	4.0	7.5
cGMP (pmol/min)	624	724	1404

*ANP = Atrial Natriuretic Peptide
NEP = Neutral Endopeptidase
FE = Fractional Excretion
cGMP = Cyclic Guanosine Monophosphate

Data in Table 4 reveal that ANP alone causes increases in sodium and volume excretion but these increments are dramatically increased when ANP catabolism is prevented by inhibition of neutral endopeptidase. Moreover, these increases occur in concert with increases in fractional excretion of ANP and in excretion of cyclic GMP, offering strong evidence that the effect is due to increased delivery of ANP to more distal sites of the nephron where it exerts a pharmacologic effect.

Though imipenem is not a peptide or protein, it is a good illustration of the potential utility of inhibiting catabolism of a compound at proximal nephron sites. Imipenem is both filtered and secreted by the kidney but it is then catabolized by a dipeptidase at the brush border of the proximal tubule. This process affects both filtered and secretory components of imipenem that appear in the nephron. This catabolism precludes any antibacterial affect of this drug at more distal sites of the kidney or in the bladder. Thus, imipenem by itself is not effective for urinary tract infections. However, if imipenem is administered with an inhibitor of dipeptidase, the amount of imipenem appearing in the urine increases from about 5% to as much as 70% of the administered dose which is sufficient to have therapeutic activity in the urine (7). The dipeptidase inhibitor used with imipenem is cilastatin, which was chosen among several possibilities simply because its pharmacokinetic profile parallels that of imipenem itself.

Data with ANP and with imipenem illustrate that inhibition of catabolism of compounds and proteins can result in a pharmacologic effect at more distal sites in the nephron. The potential clinical utility of this phenomenon with imipenem is obvious in that it allows use of this drug for treatment of urinary tract infections. The utility of such an approach with ANP has not been determined through clinical studies but might be important in enhancing the effects of ANP in clinical conditions in which circulating concentrations of ANP are elevated (for example, congestive heart failure), but in which little natriuretic effect occurs that can be attributed to ANP. Inhibition of its catabolism at the proximal nephron might allow endogenous amounts of ANP to exert a natriuretic effect and therefore be beneficial in the clinical conditions in which ANP itself is elevated. Whether or not such a strategy could be clinically useful awaits further studies.

124

CONCLUSION

The kidney plays vital roles in the elimination of proteins and peptides. Understanding these roles allows their manipulation in a fashion that will allow use of proteins and peptides more effectively as therapeutic agents. Promising advances have been made in this field and it is anticipated that more will follow.

REFERENCES

1. Mast AE, Salvesen G, Schnebli H, Pizzo SV (1990) J Lab Clin Med 116:58-65

2. Beauchamp CO, Gonias SL, Menapace DP, Pizzo SV (1983) Bioanal Biochem 131:25-33

3. Bayati A, Källskog Ö, Odlind B, Wolgast M (1988) Acta Physiol Scand 134:65-74

4. Hallewell RA, Laria I, Tabrizi A, Carlin G, Getzoff ED, Tainer JA, Cousens LS, Mullenbach GT (1989) J Biol Chem 264:5260-5268

5. Ogino T, Inoue M, Ando Y, Awai M, Maeda H, Morino Y (1988) Int J Peptide Prot Res 32:153-159

6. Margulies KB, Cavero PG, Seymour AA, Delaney NG, Burnett JC (1990) Kidney Int 38:67-72

7. Barza M (1985) Ann Intern Med 103:552-560

Discussion : THE ROLE OF THE KIDNEY IN ELIMINATING PROTEINS AND
PEPTIDES

H.J. Röthig

Does anyone know of small peptides which may be secreted by the tubular system so that renal excretion exceeds glomerular filtration?

P. du Souich

Argine vasopressin is filtrated and also secreted into the tubule, and in fact this secretion can be inhibited.

R.G. Werner

Is there any influence of the polyethylene glycol coupling on the activity of the peptide? And, what about stability and shelf life?

D.C. Brater

As far as the loss of activity is concerned, this is highly variable, and it has been shown that there are molecules where the coupling of polyethylene glycol eliminates all activity, while in others virtually none is lost, and in some the loss is intermediate. So in many cases enough activity seems to be preserved as to make them potentially therapeutically useful. On the other hand I do not know about shelf stability of a modified peptide.

M.M. Reidenberg

Is the kidney an important eliminating organ for fragments of some of these proteins that have been partially hydrolyzed, and if so might these accumulate in people with poor kidney function?

D.C. Brater

I would assume that the kidney is certainly an important organ for the elimination of these fragments. If one does not see accumulation when this process of elimination is not functioning one has to postulate the existence of alternative pathways, which are probably going to be highly variable. But I would think that one would certainly have to worry about them.

D. Maruhn

In the case of alpha 1 proteinase inhibitor we should not forget that there is an alternative, that is to produce this protein in a glycosilated form from plasma which has considerable half life and is probably acceptable for therapeutic purposes. On the other hand, as far as the saturable catabolic process in the kidney is concerned, it could be that the lysosomal capacity in the proximal tubule is what is really saturated.

D.C. Brater

In the SOD studies we also monitored some of the proximal tubule enzymes which I believe are lysosomal. And I would presume that if we were overloading the lysosomes we would have seen an increase in the levels of those enzymes in the urine, but we saw no difference in enzymuria with 45 mg/kg as opposed to 1 mg/kg. I would think that this data might be some indirect evidence that what we are really doing is overwhelming the peptidase rather than the lysosomal capacity.

J.A. Galloway

One possible example of your model is our finding that although human proinsulin is disposed of primarily in the kidney, and the half-life in patients with renal failure is prolonged, virtually no proinsulin is found in the urine.

P. du Souich

I imagine that accumulation of peptides in patients with renal failure is unlikely, in view of the large amounts of endopeptidases in other tissues, such as the lung or the intestines. Also, I have a question concerning SOD. Since it exerts its scavenger effect in the cytosol usually, how the binding to a large molecule is going to affect the entry of the superoxide dismutase into the cell?

D.C. Brater

As far as I know, in an animal model of cardiac ischemia and reperfusion injury the coupled SOD does seem to get to where it needs to go. This is just indirect proof, and of course efficacy may be by a different mechanism, but I think yours is a very good question.

L. Gauci

Since we are talking about pharmacokinetics I think that one of the major differences between protein drug development and conventional drug development is the rendering of plasma kinetics almost useless in the former.

M.M. Reidenberg

I think that too often kinetics studies are viewed as the end rather than the means. I think we need to differentiate the kinetics of the molecule from the kinetics of the effect. And that information on the kinetics of drug effect rather than just the disappearance rate of the compound probably can be useful in developing dosage regimens. Bob Meyer gave an example where kinetics of the molecule itself is important with respect to the releasing factors and whether one wants a pulsatile pattern or a steady-state pattern. So I think the issue of kinetics like any other scientific methodology is related to what is the question that the methodology is intended to answer rather than saying the methodology is an end into itself. The methodology development was an end several decades ago when the whole field of pharmacokinetics itself was being developed but now that the methods exist it is up to us to use them to answer questions rather than as ends in themselves.

EVALUATION OF BIOTECHNOLOGY PRODUCTS

© 1991 Elsevier Science Publishers B.V. (Biomedical Division)
The clinical pharmacology of biotechnology products.
M.M. Reidenberg, editor

POSSIBLE PROBLEMS ASSOCIATED WITH CYTOKINE CONTAMINATION OF BIOTECHNOLOGY PRODUCTS

ANDREW GEARING*, MEENU WADHWA & ROBIN THORPE

*British Biotechnology, Watlington road, Cowley, Oxford, U.K. and NIBSC, Blanche Lane, South Mimms, Potters Bar, Herts, U.K.

INTRODUCTION
        Animal cells have been used in the production of biological medicines for many years. Routine control testing of these products has emphasised elimination of contaminants such as host cell proteins in immunogenic quantities, host cell nucleic acids, micro-organisms, viruses or pyrogens (1). It has become clear in recent years that animal cells can produce a range of potent biologically active mediators known as cytokines (2). Cytokines are known to cause profound local and systemic inflammation, and to affect the function of most cell types. In view of this we undertook to investigate the production of cytokines by a number of cell lines commonly used to produce biological medicines, and also assayed their levels in several final products. This paper reviews our findings and assesses the potential risks and ways to minimise them.

MATERIALS AND METHODS
Cell lines.   The cell lines were grown as indicated or as previously described (3). Supernatants were harvested from the lines by centrifugation either from confluent cultures of adherent lines or at the same time that product harvest would normally occur for suspension cultures. Ascitic fluid was also prepared from some of the hybridoma lines grown as ascitic tumours in mice.

Stimulation of cell lines.   Lines were stimulated with $IL1\alpha$ 10IU/ml, LPS 10ng/ml or by addition of a number of wild type or vaccine strain viruses.

Cytokine assays.   IL1 was measured using the NOB-1 bioassay which detects $IL1\alpha$ and $IL1\beta$ with a sensitivity of 500fg/ml (4). IL6 was measured using the B9 bioassay which can detect 1pg/ml (5). $TNF\alpha$ and $TNF\beta$ were measured using the L929 bioassay which can detect 10pg/ml (6). GCSF and GMCSF were detected using two-site immunoassays (7)(Insight GM ELISA, MRL, Australia). Cytokine standards from NIBSC were used to calibrate the assays. 1 unit of IL1 corresponds to approximately 10pg, 1 unit of TNF to 25pg, 1 unit of IL6 to 200pg, 1 unit of GCSF to 10pg and 1 unit of GMCSF to 100pg of recombinant proteins.

Vaccines and Recombinant Protein products. These were made up to their final dosage volume, or to 1mg protein/ml, prior to assay for cytokine levels.

RESULTS
Our initial experiments compared the levels of IL1, TNF, and IL6 in the culture supernatants of a variety of cell lines producing monoclonal antibodies (Table 1). Two were Epstein Barr virus-transformed human B lymphoblastoid lines, two were Heterohybridomas of EBV-transformed human B cells with a murine myeloma line, and ten were murine hybridoma cell lines. Both lymphoblastoid lines produced IL1 and TNF but not IL6 and one of the murine hybridomas produced IL6. None of the other lines produced detectable levels of any of the cytokines.

TABLE 1.
CYTOKINE LEVELS IN SUPERNATANTS FROM CELL LINES PRODUCING MONOCLONAL ANTIBODIES.

CELL LINE	IL1	TNF	IL6
EBV	7, 1	125, 16	0,0
HETEROHYBRIDOMA	0,0	0,0	0,0
HYBRIDOMA	0,0,0,0,0,0,0, 0,0,0,	0,0,0,0,0,0,0, 00,0	0,0,0,0,0,0,0, 0,0,2

All values are in U/ml, no GCSF or GMCSF could be detected in the EBV transformed lymphoblastoid lines or heterohybridomas.

We also compared the levels of cytokines in culture supernatants of the same murine hybridomas grow as ascitic fluid (Table 2). Nearly all of the ascitic fluid samples contained IL1, TNF and IL6. The highest IL6 levels were seen in the ascitic fluid of the line which made IL6 in culture.
A number of epithelial and fibroblast cell lines commonly used for the production of viral vaccines, or in the expression of recombinant proteins were cultured in IL1 or bacterial endotoxin (LPS) to determine their potential for cytokine release (Table 3). Most lines produced low levels of cytokines when unstimulated. The exceptions were CHO cells which made 3.7 U/ml IL6, Vero cells 11

U/ml of GCSF, and HELA cells 29 U/ml TNF. CHO cells constituitively release IL6, on some occasions up to 100 U/ml. Following stimulation with IL1 most lines increased production of IL6 and GCSF. MRC5 cells were particularly sensitive to IL1, secreting 100 U/ml of IL6 and 5760 U/ml of GCSF. LPS had negligible effects on most of the cell lines.

TABLE 2.
CYTOKINE LEVELS IN ASCITIC FLUID OF MURINE HYBRIDOMAS

HYBRIDOMA	IL1	TNF	IL6
1	1	10	2
2	3	9	1
3	1	3	1
4	1	1	2
5	1	0	1
6	0	1	2
7	0	4	0
8	1	1	0
9	1	6	15
10	2	6	1

All values in U/ml
Hybridomas used are same as in Table 1.

When MRC5, Vero and HEP2 cells were stimulated with a variety of virus strains a complex pattern of cytokine release was seen (Table 4) MRC5 cells could produce extremely high (ug/ml) levels of IL6 and GCSF in response to flu or mumps, and high levels of TNF in response to polio. Vero cells made high levels of IL1 and TNF in response to TypeII polio and High levels of IL6 in response to all polio types. HEP 2 cells made high levels of TNF in response to Type I and II polio and IL1 in response to Type II polio.

TABLE 3.
CYTOKINE PRODUCTION BY ENDOTHELIAL AND FIBROBLAST CELL LINES

CELL	STIMULUS	IL6	IL1	TNF	GCSF	GMCSF
MRC5	MEDIUM	0.1	ND	1.8	2.6	ND
	LPS	0.1	ND	0.4	2.6	ND
	IL1	100.2		0.3	5760.0	2.5
HEP2C	MEDIUM	0.2	0.1	ND	ND	ND
	LPS	0.4	0.1	ND	ND	ND
	IL1	0.3		ND	3.0	ND
HELA	MEDIUM	0.1	ND	29.0	ND	ND
	LPS	0.1	ND	3.2	1.6	ND
	IL1	0.3		44.0	2.1	ND
VERO	MEDIUM	1.2	ND	ND	11.0	ND
	LPS	6.3	0.1	0.3	26.0	ND
	IL1	6.0		ND	24.0	ND
CHO	MEDIUM	3.7	ND	ND	ND	
	LPS	4.3	ND	ND	ND	
	IL1	4.3		ND	50	

TABLE 4. STIMULATION OF CYTOKINE RELEASE BY VIRUSES

CYTO-KINE	CELL	FLU	POLIO I	POLIO II	POLIO III	MEAS	M V	M W	RUBELLA
IL6	VERO	45	2000	2750	3000	60	2	85	23
	HEP	23	40	30	12	1	1	1	1
	MRC5	35000	30	275	125	300	2000	10000	5
GCSF	VERO	80	ND	100	140	100	80	140	120
	HEP	1300	140	ND	ND	120	ND	100	280
	MRC5	10000	200	600	220	200	300	3500	350
IL1	VERO	3	12	120	37	6	1	9	ND
	HEP	2	1	70	1	6	2	5	2
	MRC5	4	ND	ND	ND	ND	ND	1	ND
TNF	VERO	ND	14	240	ND	5	6	3	6
	HEP	1	280	120	2	1	ND	ND	ND
	MRC5	3	144	144	29	ND	ND	ND	ND

LEVELS OF CYTOKINES ARE IN U/ML. ND = NOT DETECTABLE
Meas = measles, M V mumps vaccine strain, M W mumps wild

A number of viral vaccines were also found to contain cytokines, particularly IL6 and GCSF in high levels (Table 5).

The rabies vaccines were the worst preparations studied. None of the clinical grade recombinant proteins made in CHO cells contained significant levels, although one preparation of a laboratory reagent HIV gp120 had 8 U/ml of IL6.

TABLE 5.
CYTOKINE LEVELS IN VIRAL VACCINES AND RECOMBINANT PROTEINS

SAMPLE	IL6	IL1	GCSF
POLIO a	27	ND	ND
POLIO b	ND	ND	ND
RABIES a	800	7	30000
RABIES b	46	ND	ND
RABIES c	640	ND	2000
RABIES d	13	ND	ND
MEASLES	27	ND	ND
RUBELLA	25	ND	ND
MMR	4	ND	ND
CHO HEP	0.1	ND	ND
CHO EPO	ND	ND	ND
CHO TPA	ND	ND	ND

DISCUSSION

The initial impetus for this work came from the observation that an ascitic fluid of a hybridoma secreting antibodies which neutralised human IL1a contained something which stimulated the IL1 responsive cell line NOB-1(8). This turned out to be murine IL1. We then surveyed the production of IL1, TNF and IL6 in the culture supernatant of 10 different Hybridoma lines and in the corresponding ascitic fluid. The results demonstrate that hybridomas do not produce IL1 or TNF but one of the ten could make IL6. When grown as ascitic tumours the same hybridomas caused the secretion of IL1, TNF and IL6, this is presumably a host response to the tumour line. Some therapeutic monoclonal antibodies have been produced as ascites, our results would suggest that this method carries an increased risk of cytokine contamination, although some hybridomas may still produce IL6.

We extended our survey to include some human lymphoblastoid lines and heterohybridomas making human monoclonal antibodies. The lymphoblastoid cells produced IL1 and TNF but not IL6, whereas the heterohybridomas produced none of these. In a recent report we have also shown that lymphoblastoid lines can also make Transforming growth factor B1 and soluble CD23 (a B cell stimulating cytokine) (9).

The epithelial and fibroblast lines showed differing patterns of constituitive and inducible cytokine release. Some such as MRC5 constituitively produced low levels of IL6, TNF and GCSF, others such as VERO produced high levels of GCSF, or HELA which made TNF and CHO which made IL6. Induction with IL1 markedly stimulated

MRC5 to make IL6 and GCSF, but had modest effects on the other cells. LPS had little effect. These results are in line with previous reports that LPS is not a potent stimulus for fibroblasts(10). Stimulation of cells with a range of viruses caused very high levels of cytokine release from some cell types. MRC5 cells produced 35000 units/ml(7 ug/ml)of IL6 in response to flu and 10000 units/ml (2ug/ml) in response to wild type mumps virus. Most cells also released IL1 and TNF. Van Damme and co-workers have also shown that viral stimulation of fibroblast lines causes similarly high secretion of both IL8 and MCP-1(11)(12). These are chemotactic and activating factors for granulocytes and monocytes respectively. Recent work also suggest that lines which are commonly used to express recombinant proteins for example COS can secrete soluble receptors for TNF, or CHO can secrete TGFB1.

Our work shows that many cell lines used in production of vaccines or recombinant proteins can secrete cytokines either constituitively, or in response to process components. These components can be the product itself as in the viral vaccines, or perhaps a biologically active protein which stimulates the cells directly. Other components include medium supplements such as growth factors which may also stimulate release. The extreme case would be growth as ascitic tumours in which the host animal produces cytokines as a protective response to the cell line. Cytokine content is not just a theoretical problem as we have shown that cytokines can be found in some clinically used vaccines (3).

The consequences of administering cytokines are variable, with local inflammation being the most likely problem. Some factors such as IL1 and TNF are pro-inflammatory, causing secondary release of many inflammatory cytokines. One hundred pg of IL1 injected intradermally into humans causes a marked inflammatory reaction (13). Other cytokines such as IL8 and MCP-1 are induced by IL1 or TNF and can themselves cause leucocyte infiltrations (14). Some factors such as GCSF and IL6 are not inflammatory. Systemic administration of IL1, TNF or high doses of IL6 can also be pyrogenic. The amounts of IL1, IL6 and TNF seen in most samples are however too low to cause fever in humans.

CYTOKINE PRODUCTION BY PROCESS CELLS

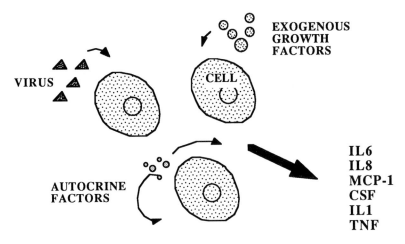

A more significant problem could arise if purification of a product caused co-purification of contaminating cytokines, doses could then reach toxic levels. A further consequence could also be the development of an antibody response against the injected cytokine. This would be more likely if the cell line used in production was non-human, and the cytokine was therefore of a different amino acid sequence. Cho cell can secrete high levels of IL6, and although this is not active on human cells, it could initiate an immune response to the hamster IL6 which could cross react with endogenous IL6. This would again only arise if some concentration of the cytokine had occurred.

Cytokines such as IL1 and IL6 may act as immunological adjuvants. This could be beneficial as in a viral vaccine but could be deleterious if antibodies to a therapeutic protein are formed. Avoidance of cytokine contamination is fortunately straightforward once it is known what cytokines are present in the process cultures. The physico-chemical properties of most cytokines are well known, and simple chromatographic procedures should remove most contaminants. An alternative approach would be to select cell lines which do not secrete cytokines as production cells. Sensitive biological, antibody and DNA probe assays are now widely available for most cytokines (15). A simple screen of process cells or samples should identify any particular cytokines and validate their removal on purification. If co-purification occurs then the purification system should be modified.

REFERENCES

1. World Health Organisation (1987) WHO Technical Report Series 747
2. Gearing AJH (1989) Thymus Update 2:127-145
3. Gearing AJH, Bird CR, Priest R, Cartwright JE, Wadhwa M, Thorpe R (1989) Lancet 1011-1012
4. Gearing AJH, Bird CR, Bristow A, Poole S, Thorpe R (1987) J. Imm Meths 99:7-11
5. Helle M, Boeije L, Aarden LA (1988) Eur J Immunol 18:1535-1540
6. Matthews N, Neale ML (1987) In: Clemens MJ, Morris AG, Gearing AJH (eds) Lymphokines and Interferons: A Practical Approach. IRL Press, Oxford, pp 221-225
7. Wadhwa M, Thorpe R, Bird CR, Gearing AJH (1990) J Immunol Meths 128:211-218
8. Gearing AJH, Leung H, Bird CR, Thorpe R (1989) Hybridoma 8:361-367
9. James K, Gardner J, Skibinski G, McCann M, Thorpe R, Gearing AJH, Gordon J (1990) Human Monoclonal Antibodies and Hybridomas In Press
10. Van Damme J, Schaafsma MR, Fibbe WE, Falkenburg JHF, Oppdenakker G, Billiau A (1989) Eur J Immunol 19:163-168
11. Van Damme J, Decock B, Conings R, Lenaerts J, Opdenakker G, Billiau A (1989) 19:1189-1194
12. Van Damme J, Decock B, Lenaerts J, Conings R, Bertini R, Mantovani A, Billiau A (1989) Eur J Immunol 19:2367-2373
13. Dowd PM, Camp RDR, Greaves MW (1988) Skin Pharmacol 1:30-37
14. Rampart M, Van Damme J, Zonnekeyn L, Herman AG (1989) Am J Pathol 135: 21-27
15. Gearing AJH, Cartwright JE, Wadhwa M (1990) In Press in Thompson AW (Ed) Immunology and Molecular Biology of Cytokines Academic Press, London

Discussion - POSSIBLE PROBLEMS ASSOCIATED WITH CYTOKINE CONTAMINATION
OF BIOTECHNOLOGY PRODUCTS

R.G. Werner

What is the detection limit in your assays?

A.J.H. Gearing

The detection limit varies from about 500 fg/ml for the IL1 bioassay up to about
400 pg/ml for the least sensitive, the GCSF assay. And that is all biologically active and
confirmed with neutralizing antibodies. So it is not a contaminant that is stimulating the
assays, it is a specific material.

A. Ganser

If one injects a cytokine, such as G-CSF or IL3, into an animal or into a patient
one has to expect the induction of a whole cascade of events including production of
IL6. Actually I wouldn't like to have IL1 in my product but if it is IL6, which I know will
induce anyway, I wouldn't think it will be too detrimental for the patient.

A.J.H. Gearing

The problem is that once you know that there is a contaminant there you should
really do something about it, and regulatory authorities would probably take that view.
If you are injecting a microgram of IL6, almost undetectable levels of IL1 and a bit of
TNF, together with a large amount of another protein the problem of adjuvanting,
becomes a possibility. IL1 or TNF can produce local inflammation, even when injected
in small quantities, so my opinion is that route of administration is important. If you give
it IV then nothing would happen. There is not enough there to cause a problem in
general.

W.M. Wardell

What you presented has considerable implications for costs of products, not only
commercial products (where I imagine the costs of what you proposed would be
absorbable) but also at the investigational level. If one were forced to absolutely avoid
any levels of cytokine contamination, what fraction would that add to the cost of the
product being prepared just for an IND? It is these costs early in the development that
add so much to the cost of the final product, through interest charges.

A.J.H. Gearing

That is an impossible question to answer, but you should take into account that most people only work with very few cell lines. The capacity of most cell lines to produce cytokines is reasonable well known. The ones that are produced in higher levels are IL6, GCSF and a few others. If the expression levels of product are reasonably high is it relatively simple to achieve purity. It is when one does ridiculous things like adding virus or using ascitic fluid or silly production schemes that problems arise.

L. Gauci

You have concentrated on the problem of contamination with cytokines when using certain types of cells. Clearly eucaryotic cells can produce many other biologically active substances, and I am thinking of hormones in particular. Do you have any feeling about whether it is really going to make the use of mammalian cells too complicated?

A.J.H. Gearing

No I don't think it is going to be a significant problem. As I said, the only thing I could see being a real problem would be minor amounts of local inflammation. I think the amounts you are going to see compared with the amount of product that comes out in the end, unless they copurify, are going to be very small.

J.A. Galloway

Shortly after we began marketing human insulin (rDNA) we received reports of three or four cases of an arthralgia-myalgia-arthritis syndrome which was associated with an increase in the erythrocyte sedimentation rate and a mild normocytic anemia. Cessation of treatment with human insulin rDNA resulted in complete resolution of the disorder. Except for those initial cases, we received no further reports related to the use of human insulin rDNA. However, we encountered a similar syndrome in the clinical trials with human proinsulin. Since none of the patients treated with human insulin or proinsulin were rechallenged, we were unable to establish a causal relationship between these agents and the condition described.

© 1991 Elsevier Science Publishers B.V. (Biomedical Division)
The clinical pharmacology of biotechnology products.
M.M. Reidenberg, editor

SELECTION OF ANIMAL SPECIES AND LENGTH OF TOXICITY STUDIES WITH RECOMBINANT PROTEINS. REVIEW OF INDUSTRY AND REGULATORS' APPROACH AND CASE HISTORY OF REC-HIRUDIN

PETER GRAEPEL, FRIEDLIEB PFANNKUCH and ROBERT HESS

Pharmaceuticals Division, Toxicology, CIBA-GEIGY Limited, P.O.Box, 4002 Basel (Switzerland)

INTRODUCTION

New chemical entities intended for use as conventional pharmaceutical products undergo toxicological testing in structured and regulated schemes. In-vitro tests and animal studies include single and daily repeated exposure. Selected rodent and non-rodent animal species are subjected to daily treatment up to the normal life-span of rodents. The experimental endpoints of toxicological testing are adverse reactions of the test systems (usually more than one test system for each endpoint) to ascending doses of the compound. The main purpose of toxicological testing is to describe those dose/exposure levels at which no damage has been seen and to base the safety assessment for a given drug on the margin of safety between the no-toxic-effect level (NOEL) and the dose levels humans may be exposed to. There are types of toxic injury which may be considered prohibitive for use of such drugs in humans. Examples are findings that indicate a carcinogenic, mutagenic or teratogenic potential. There are instances where such damage can be considered as an acceptable risk, e.g. in treatment of life-threatening disease. The benefit-risk analysis must be based on the therapeutic aim, the nature of the disease and the type of patient to be treated.

What are recombinant proteins and why are they different from ordinary new chemical entities in drug development?

Recombinant proteins

- are large molecules with complex structures, requiring a comparably disproportional effort to characterize them analytically;
- have particular genetic backgrounds (vector) and are identical or related to similar naturally occurring proteins or peptides;
- are in short supply in the early stages of development; large quantities are, however, necessary to carry out regular preclinical toxicity studies;

- are usually intended for therapeutic regimens that fundamentally differ from those of conventional drugs;
- are multifunctional molecules with a potentially wide range of pharmacodynamic effects; adverse effects produced in a biological system are usually the outcome of dosing in excess of the physiological range rather than the result of "inherent" toxicity, at least in the homologous species;
- are species-specific and are therefore likely to provoke immunological reactions when given systemically to foreign organisms;
- may reduce the biological activity of the heterologous protein because of formation of neutralizing antibodies.

The immunogenicity of recombinant products introduces the topic of this paper, i.e. the selection of species in real life. However, it is also to a large degree linked with the question of study design and duration. But before going into this discussion, the industry and regulatory procedures shall be reviewed.

PHARMACEUTICAL INDUSTRY AND REGULATIONS

There are still very few publications from authors within the industry on the question of the design of safety studies with recombinant drug products. Waife and Lasagna (1985[1]) see the regulatory pathway in the modification of existing procedures through the issuing of guidelines which can be updated as knowledge increases. We have pointed out (Teelmann et al., 1986[2]) that toxicology procedures which are incorporated in the standard guidelines for conventional drugs are inappropriate and of little value in predicting possible adverse effects of endogenous polypeptides and proteins. We developed instead a modified testing programme to demonstrate non-clinical safety rather than target-organ toxicity. In a separate publication of work done for ECETOC (1987[4]) we have assessed the available tests for use in the investigation of possible alterations in immune function (Trizio et al., 1988[3]). Cossum (1989[5]) proposed that each protein be treated on its individual (i.e. mostly pharmacological) merits and that sensible panels of toxicology procedures be created. We have also advocated a rational approach which should consider pharmacologic strategies rather than fixed testing procedures (Hess, 1990[6]). To ensure safety of a human protein or a related polypeptide it is mandatory to recognize adverse reactions induced by administering unphysiological amounts to target receptors (exaggerated pharmacological action), by effects on non-target receptors (unphysiological action), or resulting from the therapeutic regimen (chronic receptor interaction, immunogenicity). This information can only be produced by experiments designed for each individual substance with due reference to its specific properties and anticipated therapeutic use.

In the USA the testing guidelines for biologics are valid also for recombinant proteins/peptides (Crouch, 1988[7]). In general terms, the extent of recommended non-clinical animal experimentation should depend on the degree of similarity of the product to the naturally occurring equivalent. For example, in the case of monoclonal antibodies the testing requirements are reduced to systemic toxicity, pyrogenicity, cross-reactivity 'in vitro' and distribution studies. The US Food and Drug Administration's attitude is further characterized by a case-by-case strategy with great emphasis on close collaboration with the agency before deciding on and during execution of the programme. The establishment of long-term standards is thus avoided. Although the performance of suitable preclinical studies is encouraged whenever possible, the FDA has, in fact, accepted clinical approaches with ascending doses when pharmacologic/toxicologic studies are not feasible (Weissinger, 1989[8]).

The Committee for Proprietary Medicinal Products has finalized basic requirements in its 'Notes to Applicants' (1987/88[9-11]). In a similar manner, but more generalizing, not case-by-case as the US FDA, the EC proposes a classification system[11] which is based on the biochemical structure ("group") of the biotechnological product. The greater the difference between the group (to which a product is assigned a-priori) and the homologous substance, the more extensive is the scope and depth of proposed safety testing. The aim is to determine possible effects of a modification of the biochemical structure of the recombinant product. The proposed battery closely resembles established procedures. It should, however, be borne in mind that the conventional toxicology tests may be too insensitive to detect minute differences in such cases. As regards selection of animal species, a relevant recommendation[11] is made: "Where it is not possible to predict the value of particular animal species for safety testing, initial investigations may be considered in any of several rodent and non-rodent species. Where studies are expected to exceed a duration of 4 weeks, test species known to be low responders with regard to antibody production against the test substance should be considered." It appears questionable whether this approach will eventually result in predictive assessments of human risk. Later on the same Note to Applicants states[11]: "Non-physiological doses or routes of administration of the product may affect the distribution, the biological effects and toxicological profile of the product." The experience of regulatory agencies in the EC, in 1987[12] at least, was still rated insufficient to give clear guidance in these matters.

The Japanese Pharmaceutical Affairs Bureau makes little difference between the origin of new drug substances. In principle, biotechnologically derived drugs have to undergo similar if not identical safety testing as drugs that are produced by conventional methods. In their Notification (1988[13]) they have stipulated that toxicity studies should include antigenicity and pyrogen tests for impurities and require the inclusion of immunogenicity ("antigenicity") tests of the substance in conventionally fixed testing procedures.

CASE HISTORY: REC-HIRUDIN

As a practical example, a small segment of the development history of a CIBA-GEIGY product is presented and discussed. A programme of subchronic (28- and 90-day) parenteral toxicity studies with rec-hirudin had been planned and initiated in the standard rodent (rat) and non-rodent (beagle) species. In order to obtain criteria supporting the selection of the dog, we conducted an exploratory comparative immunogenicity study. The objective was to compare the immunogenic potential of rec-hirudin with that of two other proteins in three different animal species. Within this 3x3 matrix we hoped to develop a ranking order for the products and animal species as regards immune stimulation. The design of these studies are summarized in Table 1.

Table 1: Comparative study of the immunogenic potential of three proteins.

```
Compounds          (1) Rec-hirudin, antithrombotic        (2 mg/kg)
                   (2) Eglin c, elastase inhibitor       (15 mg/kg)
                   (3) Aprotinin, proteinase inhibitor  (2.8 mg/kg)

Route              Intravenous (bolus) injection

Species            Dogs (pure-bred beagle)    (n =  6)
                   Rabbits (chinchilla-type)  (n = 10)
                   Baboon (wild-caught)       (n =  2)

Parameters         Anaphylactic reactions, Skin testing, Specific
                   antibodies

Study Design:

                         Test Day
                       ┌─ -1 ─┐
       Application →    │   1  │   ← Blood sampling
                  →     │   2  │          & Skin testing
                  →     │   3  │
                  →     │   4  │
                  →     │   5  │
                  →     │   6  │
                  →     │   7  │
                  →     │   8  │
                  →     │   9  │
                  →     │  10  │
                       └─ 11 ─┘   ← Blood sampling

                       ┌─ 38 ─┐         ← Skin testing
    1st booster →       │  39  │
                       └─ 46 ─┘   ← Blood sampling

                       ┌─ 66 ─┐         ← Skin testing
    2nd booster →       │  67  │
                       └─ 74 ─┘   ← Blood sampling
```

On theoretical grounds, immunologic reactions could be expected for rec-hirudin.

Eglin c had previously caused some allergic reactions when given to volunteers in Phase I clinical studies. Development of this compound was therefore terminated and the knowledge of its true immunogenic properties in man remains incomplete. Nevertheless, because of evidence of an 'allergenic' potential we decided to include this product as a reference.

Aprotinin (Trasylol^R), a polyvalent proteinase-inhibitor, is recommended for short-term prophylaxis or therapy of acute shock
situations of various aetiologies. Although we had no animal data
for aprotinin, it is known that this compound can elicit anaphy-
lactic or anaphylactoid reactions in humans. It was therefore included as a second reference product in the evaluation.

The results of the immunogenicity tests are summarized in
Table 2.

Table 2: Results of Immunogenicity Test with Three Proteins

		rec-Hirudin	Eglin c	Aprotinin
Anaphylaxis	Dog	---	4/6	4/6
	Rabbit	---	---	---
	Baboon	---	---	---
Pos. Skin Test	Dog	---	3/6 (B2)	3/6 (B1/B2)
	Rabbit	---	---	---
	Baboon	---	---	---
Specific Anti-	Dog	3/6 +	6/6 +++	5/6 ++/+++
bodies	Rabbit	---	2/10 +	5/10 +/++
	Baboon	---	1/2 (+)	2/2 ++

B1, B2 = reaction one day prior to Booster 1 or 2
--- = no response
Level of antibody formation: + = low   ++ = medium   +++ = high

Specific antibodies (Table 3 and Fig. 1) were determined in the
plasma of dogs, against either of the three test compounds. Quantitatively, however, rec-hirudin was the least active (approx. 10 -
100 times less) at only half the incidence of eglin c and aprotinin). Rabbits and baboons did not develop specific antibodies
against rec-hirudin, but small or moderate amounts against eglin c
and aprotinin, respectively.

Table 3: Comparative immunogenicity studies: Antigen-binding capacity (μg antigen bound by 1 ml serum; error of determination estimated to be 30 %) of animal sera from tests.

Dose	rec-Hirudin 2 mg/kg		Eglin c 15 mg/kg		Aprotinin 2.8 mg/kg	
Booster	1	2	1	2	1	2
Dog	3	9	108	577	7	62
	1	9	211	740	79	227
	2	3	50	301	115	247
	–	–	12	130	8	49
	–	–	4	ND	63	108
	–	–	T	T	–	–
Rabbit	–	–	2	12	16	27
	–	–	–	T	13	18
	–	–	–	–	3	9
	–	–	–	–	2	3
	–	–	–	–	8	11
	–	–	–	–	–	–
	–	–	–	–	–	–
	–	–	–	–	–	–
	–	–	–	–	–	–
Baboon	–	–	1	T	20	51
	–	–	–	–	19	112
	All values were negative pretest and at day 11. T = Trace, ND = Not Done					

148

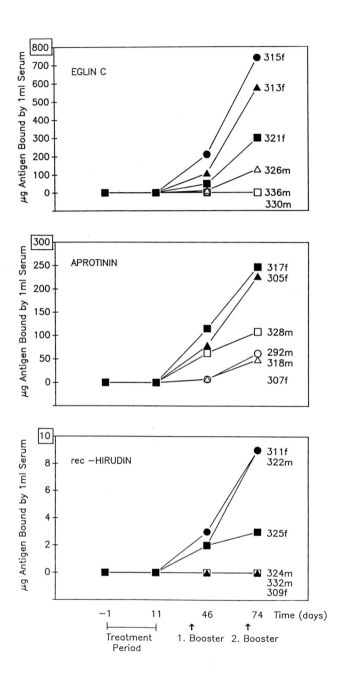

Figure 1: Comparative immunogenicity studies. Specific antibody formation over time (Note the different scales!). Species: Dog

These results confirm that the dog is an animal species particularly susceptible to immune stimulation. Rabbits and baboons appear to be much less responsive. The results also suggest that, under the conditions of these experiments, there was a quantitative difference between the immunogenic potential of the three products. It should be emphasized, however, that the extrapolation of these findings to the clinical use of these drugs is doubtful.

Given the susceptibility of the dog, we could expect a minor immunogenic potency for rec-hirudin in this species. Therefore the use of the beagle, a readily available non-rodent species for subchronic toxicity testing seemed justified, rather than the baboon.

A 28-day intravenous toxicity study in dogs at 20 mg/kg was uneventful. As in the parallel rat study, no findings were obtained that indicated any adverse properties of rec-hirudin. In particular, no evidence of immunotoxic reactions was seen. Rats which received rec-hirudin at daily subcutaneous injections of 20 mg/kg showed bleeding at the injection site but no internal haemorrhages. In the dog bleeding occurred at the intravenous injection sites.

In the course of the following 90-day i.v. study with rec-hirudin, 4 out of 18 dogs, independent of the dose administered (10 or 25 mg/kg), died or had to be killed in moribund condition after approx. 4 and 9 weeks. (Another dog survived, although the clinical condition was similar to that of the dogs which died or were killed.) Microscopically, pronounced generalized vasculitis with fibrinoid necrosis of the vessel wall and perivasculitis, involving mostly middle-sized arteries was noted in these animals, especially marked in the heart, kidney and gastrointestinal tract. Profuse internal bleeding was the terminal event, as a consequence of the anticoagulant effect of rec-hirudin combined with the vascular disease. As in the 28-day study, no effects of treatment were seen in the other 13 dogs, except some formation of specific antibodies (see Table 4).

Table 4: Rec-hirudin: Three-month intravenous toxicity study in dogs. Determination of specific antibodies (potency relative to standard serum). Test day 25, for follow-up animals also on day 115.

rec-Hirudin		10 mg/kg	25 mg/kg	
Males	Main study	36.8#	19.6	
		neg.	0.5	
		0.4	neg.	
	Follow-up group		9.1	0.5
			9.1	0.3*
			5.6	0.4
Females	Main study	19.6	4.3#	
		14.3#	0.2*	
		4.4	neg.	
	Follow-up group		0.6	0.5
			0.4	0.3*
			13.4@	

Animals killed between day 25 and day 67
@ Animal died on day 40
* Values below half maximum value of the positive control. Calculation based on 30% values (= 0.3 x (optical density of positive standard at 1/500 dilution)

Thus, the 3-month study in dogs, which was considered to be part of the package possibly required in support of the non-clinical safety evaluation of rec-hirudin, was hampered by immunotoxicity. The dog therefore proved to be an unsuitable animal for further longer-term toxicological testing of this product. On the other hand, this study confirmed that rec-hirudin is pharmacologically active in the dog as demonstrated by determination of coagulation parameters (not shown), the possible duration of the test being limited by immunogenic properties. The 90-day study in rats was uneventful.

This situation is illustrative of the uncertainty surrounding the selection of animal species and the design of toxicological studies with recombinant proteins. A number of questions arise:

- Does the occurrence, as a function of duration of treatment, of immunological adverse effects in an animal species invalidate the results obtained with shorter-term studies in this species? We propose the answer is no.

- Is the occurrence of unequivocal signs of exaggerated pharmaco-dynamic effects (bleeding in the case of rec-hirudin) enough to assess safety of a drug candidate, if the dose level eliciting them is sufficiently higher than the anticipated clinical dose in man? The answer should be yes. In this case the toxicological tests performed, e.g. with rec-hirudin in dogs up to those dose levels which, with progress of time, led to the immunologic side effects, would be considered valid. In other words, not the species per se but the circumstances would determine the usefulness of toxicological testing.
- Is qualitative toxicity, e.g. the occurrence of widespread vascu-litis in dogs, prohibitive? The answer is no if its species-spe-cifity is demonstrated (cf. Starzl et al., 1989[14]).
- How valid are tests in animals known to possess little or no res-ponsiveness to immunogenic stimulation? Certainly these animals contribute nothing to the assessment of an immunogenic potential of the test substance in humans. But, owing to large inter-species variation, neither do the susceptible species. With res-pect to the proof of adverse effects as a consequence of supra-physiological dosing, the answer should be positive.
- Since results obtained in heterologous species may be of doubtful predictive value for man, what animal should be selected for preclinical studies? The answer is to select, whenever feasible, a species which has pharmacodynamic responses similar to those anticipated in humans, and, presumably, similar pharmacokinetics. Under this premise, the choice may often be limited to non-human primates. In order to conserve precious animals, the possibility of their re-use should become an acceptable practice in safety evaluation. However, non-primate but pharmacologically responsive species (e.g. rodents) have also been useful for safety assess-ment.
- Since, conventionally, the duration of the toxicity study is re-lated to the length of the proposed clinical use, what study design is appropriate? The answer is that it should fully conform to the proposed clinical dosage regimen, unless studies are rendered unfeasible by immunogenicity, as exemplified here for rec-hirudin, or by differing pharmacokinetics in the animal. In such cases, the most feasible approach is chosen and justified.

CONCLUSION

Selection of animal species and duration of toxicity studies are only two dimensions of the multifaceted designs in safety evaluation strategies. They have to be tailor-made and are adapted to the properties of the recombinant product in development. In the design of the toxicological programme one should take into account the intended purpose of the drug product, e.g. vaccination, replacement therapy, biological response modification, specific cell- or system-targeting. In the final analysis, there may be no animal species possessing the characteristics or presenting the essential substrate necessary to demonstrate a specific pharmacodynamic principle of a novel drug product. Left with incomplete information, interpretation of findings obtained by supraphysiological dosing regimens and/or paraphysiological modes of administration in the usual toxicity tests becomes doubtful. The most important factor limiting both selection of species and duration of testing is at present the immunogenic potential of recombinant proteins and peptides.

The varying susceptibility of different animal species and the failure to predict immunotoxic phenomena lead to more trial-and-error situations with recombinant proteins than with conventional drugs. The continuation of an unbiased and open cooperation between pharmaceutical manufacturers, physicians and regulatory authorities is essential to assure the preclinical safety of the new products.

ACKNOWLEDGEMENT

We gratefully acknowledge Dr. H. Towbin (Ciba-Geigy Limited, Basel, Pharma Research) for performing the determinations of specific antibodies.

REFERENCES

1. Waife SO, Lasagna L (1985) Regulatory Toxicology and Pharmacology 5:212-224

2. Teelmann K, Hohbach C, Lehmann H, The International Working Group (1986) Arch Toxicol 59:195-200

3. Trizio D, Basketter DA, Botham PA, Graepel PH, Lambre C, Magda SJ, Pal TM, Riley AJ, Ronneberger H, Van Sittert NJ, Bontinck WJ (1988) Dd Chem Toxic. 26:527-539

4. ECETOC Monograph No. 10 (1987)

5. Cossum PA (1989) J Am Coll Tox 8:1133-1138

6. Hess R (in press) In: R. Hess (Ed.), Arzneimittel-Toxikologie, Georg Thieme, Stuttgart

7. Crouch ML (1988) Drug Res 38:947-949

8. Weissinger J (1989) Regulat Tox Pharmacol 10:255-263

9. Committee for Proprietary Medicinal Products, Commission of the European Communities (1987) Notes to Applicants for Marketing Authorizations on the Production and Quality Control of Medicinal Products Derived by Recombinant DNA Technology, III/860/86-EN, Ad Hoc Working Party on BIOTECHNOLOGY/PHARMACY, June 1987, Rev. 8 FINAL

10. Committee for Proprietary Medicinal Products, Commission of the European Communities (1987) Notes to Applicants for Marketing Authorizations on the Production and Quality Control of Monoclonal Antibodies of Murine Origin Intended for Use in Man, III/859/86-EN, Ad Hoc Working Party on BIOTECHNOLOGY/PHARMACY, June 1987, Rev. 7 FINAL

11. Committee for Proprietary Medicinal Products, Commission of the European Communities (1988) Notes to Applicants for Marketing Authorizations on the Pre-Clinical Biological Safety Testing of Medicinal Products Derived Biotechnology (and Comparable Products Derived from Chemical Synthesis), III/407/87-EN, Working Party on SAFETY OF MEDICINES, Ad Hoc Working Party on BIOTECHNOLOGY/PHARMACY, June 1988, Rev. 5 FINAL

12. Bass R, Schreibner E (1987) Arch Toxicol Supp 11:182-190

13. Japan Pharmaceutical Affairs Bureau, Ministry of Health Notification No. 1-10 (1988)

14. Starzl Th, Todo S, Fung J, Demetris AJ, Venkataramman R, Jain A (1989) Lancet 2 (8670; Oct. 28):1000-1004

Discussion - SELECTION OF ANIMAL SPECIES AND LENGTH OF TOXICITY STUDIES WITH RECOMBINANT PROTEINS. REVIEW OF INDUSTRY AND REGULATORS' APPROACH AND CASE HISTORY OF REC-HIRUDIN

H. Ronneberger

How pure was your preparation? We have carried out animal studies, up to three months in rats and monkeys, with another preparation of hirudin and we have never seen antibody formation, except in special immunization experiments, for example, immunization of guinea pigs together with Freund's adjuvant. Also, in phase I clinical studies no antibody formation was seen in normal volunteers.

P. Graepel

We used a highly purified product, the same that was used in phase I and phase II studies in humans, where we had no immunogenic reactions.

L. Gauci

I would like to disagree with some of the things that you have said. Firstly, I do not accept that it is established in a convincing way that across the xenogenic barrier one can establish models that will predict immunogenicity in man. Also your experiments were done intravenously, which is known to be the least immunogenic way of applying a protein. So I think the confusion that you have with your results is not at all surprising, and I do not believe that you are going to be able to predict what is going on in man by using such animal models. Secondly, I think the greatest need is for the toxicologist to come closer to the clinician rather than to the health regulatory authorities. The health regulatory authorities in these areas have been guided by prudence. What the clinician needs is help to understand whether it is dangerous or not to go into normal volunteers or patients. And the animal toxicology in the area of protein drugs has been virtually useless in helping the clinician to plan proper phase I studies.

P. Graepel

You are absolutely right that the IV administration mode is the least desirable in terms of immunogenicity. However, we are developing this drug precisely for IV use, and it was sort of a compromise to do it this way. I also agree with you that it is impossible to make prediction about the behavior of this type of drugs in man based on all these animal models. However, we wanted to have a ranking order between three compounds suspicious of or with proven immunogenicity, standardizing in some ways the doses and the other experimental conditions. And only for the selection of the animal toxicity studies we have done this.

L. Gauci

In my view acute toxicity is of interest and useful. However, with two week or four week toxicity studies one cannot forget that proteins reaching across the xenogenic barrier may be being treated like an abnormal protein even if antibodies cannot be measured.

A. Ganser

As a clinician working with hematopoietic growth factors, I would stress that in order to help the clinician in making a decision how to plan phase I trials, it would be very important to know the physiological action of the cytokine or protein in the animal species tested. And I think what one needs is an animal specific or the species-specific cytokine. For instance, if you do your studies in monkeys then you would need the monkey recombinant protein in order to find out the physiological action of your cytokine. Otherwise you will run into problems with the development of neutralizing antibodies against the tested cytokine. And I think this becomes especially important, if you devise phase I trials with two or more cytokines. You can never be sure what will happen in the human situation if you only have pre-clinical data with recombinant human cytokines, but not species specific cytokines.

A.J.H. Gearing

I think that is true, but also one should remember that some of the animal cytokines have different properties from the human equivalents. Thus, GM-CSF and IL3 behave differently in mice than they do in humans.

P. Tanswell

I would like to come to the support of the toxicologist. If the clinician wishes to plan a phase I trial he wants to know whether the substances are efficacious, and there he looks at pharmacological data in animals, and he wants to know whether it is safe; here the only information available to him are the results of toxicity trials. However imperfect these may be are the only data available to assess safety prior to the first administration in man. Regarding the species question, the toxicologist's hands are tied. He has to use a rodent species, that is in the guidelines, and a non-rodent species. Of the non-rodent species available he has a choice virtually of only a dog or a non-human primate

156

because a rabbit is not strictly non-rodent. So, I think we should collaborate more closely with the toxicologists and realize the uses and limitations of these studies.

A. Ganser

I think we need the toxicology studies, that is right, but to do real phase I trials one needs additional pre-clinical trials and I think that to a large extent, these should also be done with species-specific cytokines.

M.M. Reidenberg

I am a little puzzled by this. If we are talking about toxicity in terms of pyrogenicity or other pharmacodynamic effects acutely identifiable, that is one thing. But if we are talking about immunogenicity, and we seem to feel that the animal is not predictive of what will happen in man, and we know that foreign proteins have a potential for inducing an immune response, then doesn't it make more sense to pay far more attention to safety assessment in the first trial in man than in doing more and more non-predictive animal experiments that are potentially misleading either in terms of assurance of safety or assessment of risk.

P. Graepel

I totally agree with you. What we have done is we have looked at the pharmacodynamic effect, which is bleeding, and we have gone as high as it has been possible. In the rat we were restricted by the bleeding of the animals which have only about 13 ml of blood. In the dog we were not limited by the dose, but we were limited by this occurrence of four immunological events which made it unwise to go higher or to treat longer because we feel this immunological event has no relevance to man.

© 1991 Elsevier Science Publishers B.V. (Biomedical Division)
The clinical pharmacology of biotechnology products.
M.M. Reidenberg, editor

TOXICOLOGICAL DEVELOPMENT OF HEMATOPOIETIC GROWTH FACTORS

HANSJÖRG RONNEBERGER
Research Laboratories, Behringwerke AG
P.O. Box 1140, D 3550 Marburg (Fed. Rep. of Germany)

INTRODUCTION

Currently, a range of growth factors involved in the proliferation of hemopoietic cells were identified and characterized. Molecular cloned hematopoietic growth factors have now been used in many clinical trials for the therapy of life-threatening diseases.

This has provided toxicologists with an appropriate challenge to develop a reasonable strategy for safety testing and the design of toxicity studies with these products.

Factors influencing growth and differentiation of hematopoietic cells include interleukins, colony stimulating factors and erythropoietin. In most cases they are named according to the cells they stimulate which include granulocytes, macrophages, erythrocytes, megacaryocytes etc. In part they stimulate also the secretion of other cytokines demonstrating that the knowledge of the regulation of hematopoiesis is increasing in complexity. Recent studies have implicated additional factors in the control of hematopoiesis. Different factors act sequentially at different stages of the growth process. For example, the combination of GM-CSF and IL-3 leads to a synergistic increase in platelet and leukocyte counts.

GUIDELINES AND RECOMMENDATIONS

In the recent years, many proposals for meaningful testing procedures for biotechnology products were published (1, 2). Most scientists agree that routine toxicological experiments developed for safety testing of xenobiotics cannot be applied fully to recombinant drugs. However, it is known that naturally occurring human polypeptides, including hematopoietic growth factors and their recombinant counterparts may have adverse effects sometimes not detected in preclinical studies. Examples

are the fluid retention with pulmonary edema after IL-2
administration to human patients or the toxic effect of
interferons. These findings may be produced by intrinsing
toxicity - no induced by their pharmacological potential - or
by exaggerated pharmacodynamic mechanisms.

This experience led to different guidelines and
recommendations in major countries.

In the U.S. a more pragmatic approach is used allowing the
exclusion of inappropriate examinations.

In the "Cytokines and Growth Factor Pre-Pivotal Trial
Information Package" (3) studies in relevant models are
proposed necessary to assess the risk in clinical trials and to
support dose, route, frequency and duration of dosing. The
amount of animal studies can be discussed with the FDA at an
early stage of the preclinical development. This program may
include acute, subacute, chronic tests, and if conducted in
relevant species, testing on reproductive toxicity,
neurotoxicity and immunotoxicity with the active substance, as
well as with excipients or contaminating substances taking into
consideration species-specificity and immunogenicity in
non-host species limiting the relevance of many usual routine
toxicity experiments.

On the other hand, Japanese regulations for biotechnology
products are similar to the requirements of normal drugs (4).
Some routine experiments may be omitted if there are specific
reasons not to conduct them. Normally this set of safety data
has to include examinations on acute, subacute, chronic, and
reproduction toxicity, antigenicity, mutagenicity,
carcinogenicity, local tolerance, and on general pharmacology
parameters.

In the European Community the guideline for the pre-clinical
safety testing for products derived from biotechnology (5)
adopts a middle course.

The amount of studies depends on individual product
characteristics and their biochemical group. Testing is
classified into three categories:

Category I are recombinant products identical to naturally
occurring human polypeptides and proteins. Pharmacodynamic,
pharmacokinetic and some toxicological studies are required.

For category II, products closely related to the human factor with known or not verified differences of their structure, more data must be submitted including reproduction and immunotoxicity experiments.

For polypeptides and proteins distantly related or unrelated to humans more detailed testing will be necessary.

No fixed battery of studies is recommended, the usefulness of tests should be discussed with the competent regulatory authorities case-by-case.

TEST SUBSTANCES

Test substances used in preclinical toxicity studies should be manufactured and formulated identical to the product used in clinical trials. This includes cloning and expression systems, purification steps, impurity profile, and the final formulation. Significant changes in the manufacturing procedure, the production facility, product specification, or formulation may lead to additional preclinical and clinical studies because recombinant products are characterized by these factors to a great extend. This means that pivotal toxicity studies should only be performed with material from a fixed validated manufacturing process with established release specifications for the final product with known stability including compatibility with the container system (vials and stoppers).

The quality of the test substance may be influenced by the production process that should prevent and eliminate possible content of host cell or inducer contaminants or those introduced by the process, such as proteins derived from the substrate, endotoxins, DNA, culture medium and viruses. Therefore, established and relevant tests for bulk material and the final container are very important.

These analytical tests have to be supplemented by pharmacological quality control assays, such as for abnormal toxicity and pyrogenicity. These additional tests cannot however take the place of a toxicological test as they involve normally only small numbers of animals which receive only one standard dose.

Excipients, diluents, preservatives etc., chosen for

formulation, must be compatible with the active substance. Diluents used in toxicity studies should be identical as those intended for marketing of the factor.

TEST STRATEGY

Toxicity studies in animals should define the toxic potential of the drug, as far as possible. However, with recombinant products there are no guarantees that these experiments will generate always relevant information. But some of these gene products are capable of producing toxic effects as severe as xenobiotic drugs. Therefore, a careful evaluation of their toxic potential in animals prior to clinical use is necessary taking into account special problems which may arise with these substances. Target organs of human toxicity may sometimes - but not always - be predicted on the basis of animal studies.

Considering known guidelines and recommendations and balancing a more scientific approach against regulatory positions, a strategy of preclinical tests for hematopoietic growth factors must be developed to satisfy both the regulators and the clinical investigators.

The design of studies should answer scientific questions and reflect the intended application with correlation to the pharmacological response which may be linear or follow a bell-shaped curve.

Immunogenicity of these human polypeptides in animals may represent a major problem to meaningful toxicological evaluation in other than studies with a limited number of applications. Antibody formation may be induced even in closely related species, such as non-human primates. This effect could be demonstrated in Cynomolgus monkeys in a one-month subacute study with GM-CSF. Therefore, the evaluation of antibody formation should be evaluated in studies of longer duration. However, it is noteworthy to state that an antibody response may also be seen in humans receiving human proteins which may be related to genetic variability. The relevance of these antibodies for possible toxic effects during long-term therapy is not clear at the moment. Further research is needed and this issue will remain a major topic in the future.

As one of the first steps of preclinical development it would be desirable to perform pharmacokinetic and screening pharmacodynamic experiments to serve as the basis for the choice of species for toxicity studies.

Sometimes, however, comprehensive ADME (adsorption, distribution, metabolism, excretion) studies are possible only to a limited extend due to the very small amounts injected. Therefore, sensitive methods for the measurement of blood levels must be developed. Such findings will also give useful information for dosage selection, duration of longer-term toxicity studies, and comparison to data gained during clinical trials.

Because some hematopoietic growth factors tend to be species-specific, well-founded animal models must be selected based on the findings of these preceding pharmacokinetic and pharmacological studies. Therefore, initial investigations in several rodent and non-rodent species should be conducted to demonstrate presence or absence of the desired pharmacodynamic response. This approach is difficult to perform with factors where the physiological effect in man is unknown.

Some hematopoietic growth factors exhibit a pattern of species cross-reactivity but are most efficient to stimulate the target cells of the same species. Human GM-CSF is not effective in rodents and responds best to human and non-human primate progenitor cells. It stimulates canine hematopoiesis (6) but the activity on the peripheral blood cells is different to that in primates (decrease of platelets, eosinophil counts remain unchanged). In this case only monkeys are an appropriate species to perform preclinical testing. In contrast to this cytokine, the pharmacologic action of erythropoietin is also exhibited in lower species usually used in toxicity experiments. Therefore, also mice and rats can be used for these studies. As a general rule, it cannot be expected to observe toxicity relevant to man when the factor is not efficacious in the animal model selected. One of the most important aims of the toxicological experiments is to demonstrate effects when the physiological response is pushed to an extreme. Some companies conduct pharmaco-toxicological studies in monkeys first, the animal species closest to man,

lower species are used if the pharmacological response is
similar to those of the non-human primate or of humans. But
primates should only be selected if clearly needed and this
should be scientifically justified. But because the monkey is
closest to man, the subhuman primate is - in some cases - the
only relevant model available for testing.

In the experiments the same route of administration as
intended to be used in man should be used, usually the
intravenous and subcutaneous routes.

The selection of preclinical doses is sometimes difficult
because during clinical trials these may change from the
anticipated clinical dose. If possible also dosages should be
used that induce toxic findings in the animals, therefore, a
multifold (at least 100-fold) of the first anticipated human
dose may be determined. If a pharmacological effect can be
measured in the animals this can also form a basis for dose
selection in the toxicity studies.

A flexible strategy is therefore required which must take
into consideration indication in the human patient,
life-threatening or not, intended use, administration route,
dosages, frequency of application, and duration of the therapy,
and the pharmacological profile of the factor case-by-case.

On the other hand, over a period of many years much
experience has been obtained in testing the safety of
biological drugs, e.g. human plasma proteins. This has led to
the development of safe drugs and forms a good foundation on
which to base appropriate tests for these new recombinant
preparations. This experience has been supplemented by the
results of toxicity studies with many recombinant drugs during
the last years.

TEST PROGRAM

In the acute studies with single administration, by the route
intended to be used in man, parameters studied are clinical
observations, body weight and autopsy findings, supplemented by
selected laboratory tests.

The repeated-dose toxicity experiments are usually carried
out in a relevant rodent and/or a non-rodent species. The
dosages employed are normally the single human dose and

multiples of this dose, administered by the route intended for clinical use. The duration is usually to last for as long as requested by the clinical application, or up to such a time as an immunological response occurs which makes continuation of the study inappropriate. With hematopoietic factors the normal duration is 2-4 weeks eventually followed by a recovery period with some of the animals. Endpoints to be studied and autopsy and histological examinations are chosen on the basis of current guidelines for standard toxicity studies. In addition pharmacokinetic and immunological tests should be carried out.

The test on local tolerance at the injection site is performed by the usual methods used for other drugs.

Some of the most important experiments are the pharmacological safety tests - i.e. the effect of the factors on physiological body functions and organ systems. These should include the cardiovascular and the respiratory systems to exclude the possible effect of vasoactive substances by partial degradation of the polypeptides in relevant animal models.

In the majority of cases tests for mutagenicity, teratogenicity and carcinogenicity are not necessary. Such studies should be included in justified cases only.

There is no clear rationale for the conduct of mutagenicity tests on naturally occuring peptides or those produced by recombinant DNA technology. No reports are available that hematopoietic growth factors may interact with the DNA. For regulatory purposes perhaps some in vitro tests in mammalian cells could be selected.

It is unlikely that meaningful or realistic data could be obtained from experiments on the carcinogenic potential of these factors. It cannot be excluded that growth factors may have a tumor promoting potential leading to a possible risk especially after therapy with known mutagens or carcinogens (e.g. cytostatic drugs). But at the moment no validated models exist to exclude this possibility.

Selected studies on possible reproduction toxicity should be performed when relevant animal models exist or when factors may have additional functions on the generative organs. For example CSF-1 uterus organ levels are increased during pregnancy and effects on implantation of the ovum has been detected in mouse experiments (7).

TOXICITY STUDIES WITH r GM-CSF AND r EPO

The mainly species-specific recombinant human GM-CSF stimulates granulocytes and macrophages after cancer chemotherapy, bone marrow transplantation, radiation etc. Acute and 30-day subacute studies were performed in Cynomolgus monkeys, single-dose experiments also in the non-relevant rabbit. In vitro mutagenicity tests included two studies in mammalian cells. Examinations of pharmacological safety endpoints were conducted in Cynomolgus monkeys. The influence of low and high doses on circulation, respiration, hematology, coagulation and serum chemistry parameters was studied. Additional experiments included tests on antibody formation in different species, on pyrogenicity and on the absence of E. coli proteins from the cell culture process.

Recombinant human erythropoietin is a non species-specific glycoprotein from murine fibroblasts. Main indication is the therapy of anemia due to chronic renal failure. Acute studies were performed in mice, rats and rabbits, subacute 30-day experiments in rats and Cynomolgus monkeys. Reproduction studies were conducted in rats (segments I, II, III). In vitro mutagenicity tests, studies on local tolerance, absence of cell culture proteins and antigenicity supplemented the safety data set of this product. Safety pharmacological experiments in monkeys and rats excluded the adverse effect on vital organ functions.

CONCLUSION

Animal studies for the assessment of the toxicological profile of hematopoietic growth factors are only meaningful if they provide relevant information to their use in humans. It requires detailed information on their structure, their pharmacodynamic and pharmacokinetic potential to define the risk for adverse effects in preclinical experiments.

Toxicity testing should not follow rigid guidelines, it must be adapted to the properties of these factors case-by-case. What tests should be conducted must be decided pragmatically.

REFERENCES

1. Zbinden G (1987) In Graham (CE (ed) Progress in Clinical and Biological Research 235. Alan R Liss, New York, pp 143-159

2. Lewis, HB (1987) In Holcenberg JS, Winkelhake JL (eds) The Pharmacology and Toxicology of Proteins, Alan R Liss, New York, pp 173-184

3. Cytokine and Growth Factor Pre-Pivotal Trial Information Package (1990) FDA, Center for Biologics Evaluation and Research

4. Data Requirement for Registration of Drugs Manufactured Using Recombinant DNA Techniques (1984) Ministry of Health and Welfare of Japan

5. EC Commission: Notes to Applicants for Marketing Authorizations on the Pre-Clinical Biological Safety Testing of Medicinal Products Derived from Biotechnology (1988)

6. Schuening FG, Storb R, Goehle S, Nash R, Graham TC, Appelbaum FR, Hackmann R, Sandmaier BM, Urdal DL (1989) Exp Hematol 17: 889-894

7. Pollard J (1990) Paper presented at the Second FDA Cytokine Growth Factor Workshop Bethesda, USA, transcript FDA

Discussion - TOXICOLOGICAL DEVELOPMENT OF HEMATOPOIETIC GROWTH
FACTORS

J.A. Galloway

I would like to make a comment about immunogenicity and antigenicity. I think that all of us would like to avoid this in our recombinant products. On the other hand, antibodies are not necessarily bad. Immunotherapy is an accepted intervention for the treatment of allergic diseases in man. In addition in the University Group Diabetes Program where patients received beef insulin for ten years and developed antibodies, they showed no long therm effect ascribable to those antibodies.

H. Ronneberger

The problem is that different regulatory agencies may react in different ways to the same problem. Some may be quite pragmatic (e.g. the F.D.A.), but some may show overconcern. For instance, in the case of GM-CSF, it was suggested that one would induce an AIDS-like state by the administration of small amounts of antigenic substances over a longer period

A.J.H. Gearing

Can I comment also that for many cytokines there are naturally existing antibodies.

R.G. Werner

I would like to make a comment on changes in production process. A change in the manufacturing process should only be considered to be significant if there is a change in product quality and the product does not meet the specifications, we have right now a real large number of quality control methods for proteins and also for impurities and I think we will be able to detect changes of product quality after a change of a production process.

You mentioned that there are natural variants of hirudin and that there are also second generation hirudins. Is there any improvement in specificity or activity compared to the natural compound?

H. Ronneberger

What is a minor change and what is an essential change you can discuss it with the authorities and if you have, for example, another purification step, of course, this is a very essential change of your product; your product is characterized by the production procedure. Perhaps it is nonsense, but nobody knows it exactly. This is always difficult to decide.

M.M. Reidenberg

As a somewhat disinterested observer to this, it seems to me that what these studies are doing is really establishing safety of the regulators and of the company rather than of the product. It seems as if we are asking the regulators what to do, as if they and we are not part of the single scientific community really investigating a brand new classification of drugs and trying to figure out what to do. And I would wonder if there is any possibility of accumulating this experience over these next few years in a systematic way so that one can review the subject in two years time or three years time and contribute to determining what should be done based on experience and knowledge so that a meeting such as this in 1993 wouldn't end up with the same kind of discussion and no scientific advance.

H. Ronneberger

Most authorities, I think, have no real experience with these products, and they want to be on the safe side and therefore they ask for these experiments and the companies have to do such experiments otherwise they will not get the approval. And if mutagenicity tests for monoclonal antibodies are required you have to perform these studies even if everybody knows that it is nonsense.

M.M. Reidenberg

I accept that as a state of the world right now, but what I am saying is that we need to think about how to make it better. In fields such as law, the academic lawyers write law review articles (at least in common law countries) that review cases and end up with an analysis that is then used to help advance law, as it has to deal with problems that never existed before. I would hope that in medicine we can do the same thing. We should be accumulating data now to indicate that certain things that regulators are now requiring for the safety of the regulators do not make scientific sense. But unless somebody can bring together 50 or 70 such examples where this was done and it doesn't make sense, three years from now they will continue to require the same things, because they have no scientific support for no longer requiring them. The data that we are talking about exists in company files; usually it's not made part of the peer reviewed literature and so an academic doesn't have the access to the data as things stand right now in order to write the equivalent of an academic law review article. What I am requesting or urging is that we change that so that we will be able to have science evolve and regulatory science evolve so that it will make sense in the future, because as things stand right now this identical meeting held three years from now would end up with the identical kind of complaints, the identical kind of data and no advance in the identical kind of excessive costs in development for which there is no value.

P. du Souich

By prerogative of the chairman, I would like to add the following: in Canada, the preclinical and clinical studies, and even the review process, are facilitated because industry, in agreement with the government, may create a panel of scientists to evaluate the progress of the dossier. The panel reports to the government. The members of the panel are not involved in the development of the drug so to avoid conflicts of interest, and will only act as referees to evaluate all aspects of the drug. This procedure fastens the development of the drug and its acceptance by the government.

R. Ronneberger

Yes of course I will include also the Canadian authorities to what I said about the FDA, it is very similar.

© 1991 Elsevier Science Publishers B.V. (Biomedical Division)
The clinical pharmacology of biotechnology products.
M.M. Reidenberg, editor

MEANINGFUL EVALUATION OF BIOTECHNOLOGY PRODUCTS

PER JUUL

Department of Pharmacology, The Royal Danish School of Pharmacy,

2 Universitetsparken, DK-2100 Copenhagen Ø (Denmark)

INTRODUCTION

During the last decade national and international regulatory authorities have had to meet the challenge of evaluating medicinal products derived from new biotechnological procedures, mainly through recombination of genetic information from dissimilar organisms (recombinant DNA technique, rDNA) and the fusion of dissimilar cells to form a monoclonal hybridoma that is viable and differentiated *in vitro* over an extended time (monoclonal antibodies).

The innovations caused by medicinal products derived by biotechnological processes comprise a) products of higher purity, b) production of larger quantities, c) modifications of the molecular structure to improve the therapeutic usefulness, including new indications, and d) potential possibility of cheaper production.

Numerous national, international and supranational guidelines have been prepared concerning the production and quality control of these products as well as concerning preclinical animal toxicity testing, e.g. the EC Notes for Guidance (1). However, very few recommendations have appeared pertaining to the clinical evaluation and the final estimate of the benefit/risk ratio (2, 3).

The purpose of this paper is to present a survey of the status of regulatory requirements/evaluation 1990 based upon experience with "old" biological products and "new" biotechnologicals. The products derived by biotechnological procedures shall be considered together, since their evaluation with regard to quality, safety and efficacy is in principle the same. However, some of the actual differences are evidently caused by the different techniques employed in their production: rDNA techniques involving *E. coli* and yeast or transformed cell lines resulting in polypeptides/proteins, which are physiological or modified ("clever proteins"), heterohybridoma techniques involving continuous cell lines or mouse ascitic fluid resulting in rodent immunoglobulins or the combination of techniques resulting in "designer" or "humanized" antibodies (3, 4, 5).

EXPERIENCE WITH "OLD" BIOLOGICAL PRODUCTS

Parenteral administration of biologicals have been used in human medicine for 3 centuries - the first blood transfusion (lamb to human) being performed in 1667 and the first variolation in 1717 (both experiments in children). However, "The First Therapeutic Revolution" started with the ideas of Pasteur in the second half of the eighteenth century. The first Nobel Prize in medicine in 1901 was awarded Emil von Behring for his work on serum therapy. Since then numerous biologicals derived from extracts of various animal and human organs, body fluids and microorganisms have been and are still being developed and used successfully in the prophylaxis and treatment of several diseases: hormones (e.g. insulin (1922), glucagon and human growth hormone), coagulation factors, vaccines, antisera (antitoxins), albumin, gamma-globulins, anti-digitalis Fab, human anti-D immunoglobulin, plasma, antibiotics, cytostatics, enzymes (hyaluronidase, streptokinase, urokinase, anisoylated plasminogen-streptokinase activator complex) *et cetera*. Many of these products were developed before the existence of regulatory authorities and professional/ethical requirements concerning clinical trials, and the route from idea to clinical use was often very short, the first clinical use frequently arising from compassionate use. These biologicals revolutionized the treatment of numerous infectious diseases, diabetes, haemophilia, acute myocardial infarction *et cetera*. Subsequent improvements of the production, purification and control resulted in medicinal products of a reasonable or high standard, e.g. insulin.

However, the biologicals are not devoid of side effects, a.o. arising from their way of production. We have had to deal with the elimination of pyrogens and microbiological contamination, and among the side effects acute and delayed hypersensitivity reactions have been and are of great concern. An illustration of these problems is, that out of 23 deaths in children reported to the Danish Adverse Reactions Committee 1968-88 eight were attributed to vaccines and 4 to allergens used in desensitization (Andersen et al., in preparation). Among the wellknown problems are the possible transfer of hepatitis virus, HIV, bovine spongiform encephalopathy and Creutzfeldt-Jakob's disease. Recently we have been faced with the eosinophilia-myalgia syndrome associated with the use of tryptophan probably caused by an impurity from the production involving fermentation using *Bacillus amyloliquefaciens* - despite a purity of more than 99.6 % (6). Some of these recent, unexpected side effects have added to the regulatory concern when dealing with the new biotechnological products.

EXPERIENCE WITH NEW BIOTECHNOLOGICAL PRODUCTS

We have had experience with new biotechnologicals for only one decade, although their development started with the elucidation of the DNA structure by Watson & Crick. During this period a number of products have appeared, some of which are already marketed, others in clinical investigation, e.g. insulin, glucagon, human growth hormone, interferons, interleukines, recombinant tissue plasminogen activator, erythropoietin, granulocyte/macrophage colony stimulating factor, granulocyte colony stimulating factor, hepatitis B vaccine, coagulation factors, murine OKT3, monoclonal antibodies specific for Gram negative bacterial lipopolysaccharide, and monoclonal antibodies alone or conjugated with radionuclides, plant and microbial toxins and oncolytic agents.

The use of these medicinal products comprises single dose administration for diagnostic purpose, short term treatment (bacterial infections, hypoglycaemia, thrombolysis, cancer, myelosuppression), long-term treatment at physiological doses (diabetes mellitus, pituitary nanism) and long-term treatment at supraphysiological doses (Turner's syndrome).

Some of these products are "genuine" human molecules, others are modified ("clever proteins"), and still others are murine antibodies or "humanized" antibodies.

Some products are rather small polypeptides (glucagon with 29 amino acids), others are large (erythropoietin with more than 500 amino acids), and the monoclonal antibodies may be even more complicated.

The products are derived from very different technologies comprising rDNA techniques involving *Escherichia coli*, *Saccharomyces cerevisiae* and transformed cell lines, heterohybridoma continuous cell lines (cell cultures) and mouse ascitic fluid.

The therapeutic benefits from the new biotechnology products have been obvious.The development of rDNA insulin may not be called a revolution, but the haematopoietic growth factors and interleukine-2 are examples of important new drugs. The higher purity with a diminished risk of microbiological contamination is an evident benefit, e.g. rDNA human growth hormone without the risk of transferring Creutzfeldt-Jakob's disease. Products being developed for use in cancer, viral diseases, parasitic diseases, autoimmune diseases and haemophilia may hopefully deserve the designation "The Second Biological Therapy Revolution."

The drugs developed by biotechnological methods are not without side effects, some of which may be potentially serious. However, so far they have mainly been caused by the active molecules and not by unexpected impurities or contaminants from the production.

Allergic/immunological reactions were to be expected when using non-human species specific polypeptides, e.g. murine antibodies. However, the use of monoclonal antibody affinity isolated factor VIII despite the appearance in the final preparation of trace amounts (ng) of mouse antibody leaching from the solid phase support does not result in clinical immunological reactions in the recipients (7). Clinical symptoms related to hypersensitivity or anaphylaxis have not been observed from the use of human peptides developed through rDNA techniques involving *E. coli* or yeast. The previously used rDNA methionyl-human growth hormone gave rise to antibodies, but these were of little or no clinical significance.

Microbiological contamination has not been a clinical problem so far.

Another serious problem is the possibility of an oncogenic potential. At present it cannot be excluded, that there may be a causal relation between the use of human growth hormone (pituitary extract or rDNA synthetized) and the development of leukaemia, although the number of cases reported so far is very small.

Despite this positive experience, all the problems mentioned rightfully are causes for concern.

EVALUATION OF BIOTECHNOLOGY PRODUCTS
Quality

Numerous guidelines exist on the production and quality control of biotechnology products including validation of virus removal and inactivation procedures (1). Evaluation of the quality aspects is important concerning "ordinary" drugs, but is crucial when dealing with biotechnological products, in particular since unwanted side effects caused by impurities and/or microbiological contamination may be very difficult, sometimes impossible to detect in preclinical animal toxicity studies and in short-term human clinical trials.

Potential for mutation drift or genetic alteration of the cloned rDNA sequence, and subtle changes of the host cells during propagation are of concern. One of the consequences may be the production of neoantigens. Evidently a thorough characterization of the final product is crucial. A number of chemical and physical tests provide information on the primary, secondary and tertiary structure of a polypeptide/protein. Although it is still difficult to obtain complete information on the tertiary structure and increasingly problematic, when the size of the protein becomes larger, these tests give reasonable assurance, that any mutant product will not escape detection. The final product furthermore must demonstrate its biological activity in appropriate *in vivo* and *in vitro* bioassays.

Purities of 95 % or more can be achieved. The impurities may be product-related substances, foreign antigens, DNA fragments, pyrogens, "excipients" used during production or in the final product, and contaminating microorganisms.

Eliminating *product-related substances* to ppm levels is difficult and normally not necessary. Individual impurities of the order of 0.1 % may be identified and quantified.

Foreign antigens, e.g. arising from the use of *E. coli* or yeast may be reduced to less than 10 ppm.

The use of mammalian cells for production requires tumour cell lines and viral vectors. Therefore, the cloned cells contain oncogens and transforming and mitogenic products. *DNA contamination* may be reduced to even lower levels than the antigens. This is of particular importance, since oncogenic effects cannot be excluded following accumulation in the recipient during chronic use. Levels of the order of 0.1 pg of specific (oncogenic) cellular DNA per mg of the drug, however, are assumed to be safe. Using mammalian tumour cells for the production of drugs, the potential risk of contamination of transforming proteins, including oncogene coded products and growth factors, must be considered. Mammalian cell lines require a viral transfecting gene segment as part of the expression vector, which may have the potential to transform cells of the recipient.

Pyrogenic reactions may be caused by pyrogens in the ordinary sense or by an effect of the actual drug substance, e.g. interferon.

"Excipients" used during the production or in the final product in particular comprise human and bovine albumin carrying the risk of viral contamination.

Products developed by biotechnological methods may carry *viral agents* pathogenic for man, which could even be tumourigenic. This problem has been recognized for many years, e.g. concerning hepatitis B and HIV in the "old" biologicals. However, Creutzfeldt-Jakob's disease and bovine spongiform encephalopathy being recognized as "slow virus" diseases were unexpected findings. Other, presently unknown viral agents cannot be excluded. The problem is of particular concern, when continuous mammalian cell lines are used for the production. Certain proteins/polypeptides are produced in mammalian cell lines (e.g. interferon derived from the Namalwa human lymphoblastoid cell line). The regulatory problem is, that some proteins, e.g. human growth hormone, can be produced by rDNA techniques using *E. coli* as well as by mammalian cell lines. The potential risk is greater using the latter technique, and the hormone may be administered for many years to children. So far, most countries as well as WHO consider products derived from continuous mammalian cell lines acceptable from a safety point of view, but the problem is still under debate.

The above mentioned examples of problems arising from the manufacturing process concern the "quality" of the final product. However, the various impurities mentioned *are* present in the final product, although in infinitesimal quantities, and the potential risk of contamination by hitherto unknown viral agents *cannot* be excluded. The benefit/risk evaluation thus is a toxicological and clinical pharmacological matter, which at present has to be considered case by case. It seems reasonable to assume, that the potential risks are presently overemphasized, but only further clinical experience including postmarketing surveillance may give an answer.

Preclinical animal pharmacology and toxicology

Guidelines on animal pharmacology and toxicology follow the principles known from "ordinary" drugs, however modified due to the nature of species specific proteins developed by biotechnology (1). If the total structure of the products could be definitely identified as identical to the biological molecule, and if impurities did not exist in the final product, animal experiments were not necessary. Since this is not the case, *in vitro* and *in vivo* bioassays as well as pharmacological experiments and limited (acute and subacute) animal toxicity testing are necessary as well as testing for local tolerance and probably also for mutagenicity. These principles evidently apply also to modified molecules and completely new biologicals. The interpretation is often very difficult, since the species specificity may preclude evaluation of pharmacological/toxicological effects, neutralizing antibodies may add to the problems, and immunological reactions may preclude long-term administration, e.g. in normal carcinogenicity testing.

At present the predictive value of these animal experiments is unknown, and the same concerns the *in vitro* tests. However, with the increasing knowledge of particularly immunotoxicology of ordinary drugs and chemicals it seems reasonable to assume, that predictive toxicometrics also comprising biologicals shall improve (8, 9). Immunotoxicological tests (involving a.o. immunohistochemistry) are already requested in the EC guidelines on preclinical toxicity testing of "ordinary" drugs - attention should be paid to possible interference with the immune system in subacute and chronic animal toxicity studies. These tests are relevant also pertaining to drugs developed by biotechnology. Whether testing of non-immunological toxicity in immuno-incompetent animals is possible avoiding the problems of species specificity, remains to be established. Studies in T-cell deficient athymic animals as well as in Severe Combined Immune Deficiency animals (SCID) might be a solution.

Clinical documentation

The usual pharmacokinetic investigations comprising absorption, distribution and disposition should be performed following single and repeated administration. These studies add to the confirmation of structural identity (e.g. concerning "generic products"), to improved routes of administration and dosage regimens (10) and to the determination of possible changes in kinetics following repeated administration, e.g. increased clearance of rodent antibodies caused by anti-antibodies. The sometimes very low plasma levels of the products cause analytical problems. The use of higher doses is not always feasible, but may sometimes be used, e.g. employing the glucose clamp technique when administering high doses of insulin.

The demonstration of efficacy of biologicals derived from whatever biotechnological technique is as easy/difficult as concerning "ordinary" drugs - most often relying on randomized clinical trials. The relevant reference substance may be difficult to choose, and in particular the duration of the trials may constitute a problem. However, all these problems are wellknown to the clinicians, the clinical pharmacologists and the regulatory authorities.

The final benefit/risk evaluation, which must be performed by the regulatory authorities, as usual involves the documented clinical effects weighed against demonstrated and potential risks. The marketing of interleukine-2 is an example of such an evaluation accepting the documentation of a certain effect in metastatic renal carcinoma, but realizing the incomplete information on dosage regimens and the possible necessity of simultaneous administration of LAK-cells. The acceptance of the indication "Turner's syndrome" concerning human growth hormone is another example. The effect on the final height of the patients is still unknown, but the experience so far justifies the decision.

In some diseases it can be anticipated, that it shall be rather easy to evaluate the benefit/risk ratio - assuming that new biotechnologicals are developed for the prevention/treatment of diseases like AIDS, malaria, cancer and autoimmune disorders.

Side effects

Potential side effects caused by biotechnological products are of great concern. The side effects wellknown from "ordinary" drugs (renal and hepatic injury, myelosuppression, cardiotoxicity *et cetera*) shall be recognized through clinical trials provided a certain (high) incidence. The three major problems when dealing with biotechnological compounds comprise contamination with "slow viruses" (prions), immunological/autoimmunological reactions and oncogenic potential. Some of these unwanted effects may be very difficult to detect - partly since we may not know what to look for and hence not how to look for it, partly due to the fact, that the side effects may turn up after long-term or even very long-term use of the products and possibly with a low incidence.

The effects of viral contamination may not be observed within years after the administration, as it is the case with Creutzfeldt-Jakob's disease and bovine spongiform encephalopathy. The same problem holds true concerning the development of malignant diseases, as it is wellknown from the experience with secondary malignancies following certain cancer-chemotherapeutics. Both these areas demand postmarketing surveillance.

With regard to the possible immunological/autoimmunological reactions no guidelines exist pointing to a basic "battery" of tests to be recommended during the clinical trials. It seems reasonable to suggest, that at least a number of patients participating in the trials involving repeated administration should be thoroughly investigated with regard to humoral as well as cellular immunological reactions. Circulating antibodies (particularly IgM and IgG) including antibodies to fermentation products or anti-antibodies concerning monoclonal antibodies should be looked for as well as antinuclear antibodies (ANA). The complement system should be studied. The possible development of circulating immune complexes should be investigated. Finally, the number and distribution of B- and T-lymphocytes *et cetera* may be analyzed by means of fluorescence activated cell sorting (FACS). These analyses in combination with a thorough clinical examination may hopefully detect possible immunological reactions at an early stage of the development of new biotechnological products. So far the clinical experience with the existing products has been positive and has not given rise to particular fears. However, the development of biologicals, which we have not previously had at our disposal, modified molecules/completely new molecules, and the possible use of supraphysiological doses for long periods in large groups of patients with non-malignant diseases, call for attention.

As it has been stated: Today's theory is always at risk from tomorrow's data - leading to the risk of a "biotechnological Chernobyl disaster", i.e. a generalized assault on all biotechnological research and products caused by side effects arising from the use of a single biotechnological medicinal product (2).

POSTMARKETING SURVEILLANCE

As with any new drug there is an increasing need for postmarketing surveillance, in particular for two reasons: The often low incidence of certain side effects and the often delayed onset rendering detection before marketing impossible.

Until now the numerous attempts to establish efficient postmarketing surveillance systems have proved rather unsuccessful. Even large phase IV studies are often too small and suffer from the lack of control groups, short duration (1 year) and an often high drop out rate. Similarly the different systems built up locally, nationally and internationally are probably unsuitable for the detection of the side effects mentioned.

A reasonable approach might be to establish "registers" of some of the patient populations in question, e.g. diabetics, pituitary dwarfs, Turner's syndrome patients, HIV-positive patients *et cetera*, when effective treatment had been established. By modern communication and computer techniques it should be feasible to generate information on several "events", even the rare and delayed types. A major obstacle is the often paranoic aversion by the public and their representatives against registers and computer systems of any kind, which process "personal information." It is to be hoped, that the recently established "Pharmacovigilance Working Party" under the EC Committee for Proprietary Medicinal Products shall be able to create a postmarketing surveillance system, which can combine data from the 12 European Member States comprising a population of more than 320 million.

CLINICAL TRIALS

Clinical trials concerning biotechnologicals do not in principle differ from trials with other drugs. However, the initial phase I studies with "ordinary" drugs are often performed using small-scale, laboratory batches, and the regulatory requirements concerning quality are limited, which experience seems to justify. When initiating the first clinical trials on biotechnological products, most often involving healthy subjects, it seems reasonable, however, to insist on quality requirements close to those necessary for marketing. This is justified by the potential side effects mentioned previously comprising viral contamination, immunological/autoimmunological reactions and oncogenic potential. At present the requirements differ very much from country to country, which is unacceptable.

VARIATIONS

When dealing with "ordinary" drugs, minor variations of the production process and pharmaceutical formulation normally do not give rise to great regulatory problems. However, concerning biotechnological products, with which we have a limited experience, it seems reasonable - at least for a longer period - to perform a more thorough analysis of the possible consequences of the variations for the properties of the final product. Even minor changes in the biological production process may lead to significant changes in biological effects, purity and nature of impurities as well as risks of microbiological contamination. At the EC level it has been decided, that even minor variations shall be presented to the Committee for Proprietary Medicinal Products for approval according to Council Directive 87/22/EC.

GENERIC EQUIVALENTS

The scientific evaluation of generic products concerning "ordinary" drugs normally does not create problems. However, when dealing with the more complicated polypeptides/proteins and the different ways of biotechnological production techniques, it is a question, whether a mere demonstration of an acceptable relative bioavailability is sufficient documentation for granting a marketing authorization. At present the different national regulatory authorities seem to agree, that biotechnological "generics" should be investigated in more detail - besides the obvious quality requirements. At least a limited toxicological examination is necessary as well as a bioavailability study and a small scale clinical investigation. Provided, that the clinical documentation is publicly available through publications, repetition of the clinical trials is evidently unnecessary, unethical and represents a waste of resources.

DISCUSSION

The biotechnological evolution/revolution so far has provided us with a few important innovations, but beyond any doubt several new products already in the pipeline shall be of great therapeutic benefit. The assessment of these products by the regulatory authorities has become an unpleasant challenge, since we are dealing with potential risks, with which we have little or no experience. Although the authorities of the US, Japan and the EC endeavour to promote research, development and marketing of these potentially valuable therapeutics, it is understandable, that the requirements concerning quality, safety and efficacy are severe, as it is apparent from the various guidelines. However, it is to be hoped, that the increasing experience with biotechnologicals shall enable more "relaxed" requirements - provided, that the fear of the potential, serious side effects turns out to be unjustified. Every regulatory authority and guideline stress the importance of evaluating the new biotechnologicals case by case adjusting the requirements to the state of the art. However, guidelines have a deplorable tendency to become at least minimum requirements, and they tend to become stationary. The American "Points to Consider" in many respects are preferable to guidelines. One extreme is the first injection (1922) of insulin in a 14 years old boy only half a year after its isolation. The other extreme is the (hopefully) unnecessary delay of the marketing of new biotechnologicals through rigid regulatory requirements. The latter situation shall do more harm than benefit to the patients, and it shall result in enormous expenses to society. Some of the new biotechnologicals already marketed are extraordinarily expensive, and since it is reasonable to assume, that large patient populations may benefit from some of these drugs, their introduction may result in a great burden on the health budgets of the industrialized countries and the inaccessibility to patients in developing countries.

Cooperation between the regulatory authorities of the EC, US and Japan is necessary to reach harmonized, scientifically justified and reasonable "requirements" for future developments in this field.

SUMMARY

Based upon experience for three centuries with "old" biological products and for a decade with "new" products derived by biotechnological procedures and recognizing the rapid advances in molecular engineering of proteins/polypeptides at the same time as the increase in the knowledge of pathophysiology and immunology, it does not seem unjustified to expect a "Second Biological Therapeutics Revolution" from biotechnologicals in the prophylaxis and treatment of viral and parasitic diseases, cancer and autoimmunological disorders. However, despite increasing knowledge and experience the benefit/risk estimate is still empirical. Present guidelines rightfully use the words "flexibility", "individualized approach", "case by case", "ad hoc" et cetera to describe the requirements involved in the development and regulatory evaluation of new biotechnology products. The necessity of an individualized evaluation is caused by several factors: Differences between biotechnological production processes (rDNA techniques involving E. coli, yeast or cell lines, heterohybridoma techniques involving mouse ascitic fluid or large scale tissue culture, and genetic engineering in combination with hybridoma technique); molecular complexity of the products; dosage regimens (single dose, short/long duration of treatment, physiological/supraphysiological doses); therapeutic indications (severity of disease).

The requirements with regard to quality are and shall continue to be strict (molecular identification, impurities including antigenic contaminants and oncogenic DNA fragments, viral contamination).

The predictive value of animal experiments and *in vitro* tests shall probably increase.

Hopefully, clinical experience shall prove, that the present fears of microbiological, immunological and oncological side effects are unjustified. At present, the clinical investigations of biotechnological products apart from the obvious purpose of demonstrating therapeutic efficacy must concentrate on potential side effects relying on clinical examination, immunological investigations and postmarketing surveillance.

The final benefit/risk assessment must weigh the inadequacies of quality and safety parameters with regard to predictive value against the severity of the diseases to be treated.

Compared to "ordinary" drugs more strict requirements are necessary concerning variations of the production process, generic equivalents and clinical trials of biotechnological products.

Present guidelines should be regarded as Points to Consider and not eternal requirements.

REFERENCES

1. Notes for Guidance (published or in preparation) prepared by the Working Parties on Biotechnology/Pharmacy and Safety of the EC Committee for Proprietary Medicinal products: Production and control of medicinal products derived by recombinant DNA technology; Production and quality control of human monoclonal antibodies; Production and quality control of cytokine products derived by biotechnology processes; Validation of virus removal and inactivation procedures; Preclinical biological safety testing of medicinal products derived by biotechnology.

2. Lasagna L (1986) Clinical Testing of Products Prepared by Biotechnology. Regulatory Toxicology and Pharmacology 6:385-390

3. Boisclair A et al (eds) (1988) Biotechnology Pharmaceuticals - a Regulatory Challenge. BIRA Publications.

4. Larrick JW (1989) Potential Monoclonal Antibodies as Pharmacological Agents. Pharmacol Rev 41:539-557

5. Mayforth RD, Quintans J (1990) Designer and Catalytic Antibodies. New Engl J Med 323:173-178

6. Belongia EA et al (1990) An Investigation of the Cause of the Eosinophilia-Myalgia Syndrome Associated With Tryptophan Use. New Engl J Med 323:357-365

7. Davis HM et al (1990) Lack of Immune Response to Mouse IgG in Hemophilia A Patients Treated Chronically with Monoclate, a Monoclonal Antibody Affinity Purified Factor VIII Preparation. Thrombosis and Haemostasis 63:386-391

8. Descotes J (1988) Immunotoxicology of Drugs and Chemicals. Elsevier, Amsterdam-New York-Oxford

9. Trizio D et al (1988) Identification of Immunotoxic Effects of Chemicals and Assessment of Their Relevance to Man. Fd Chem Tox 26:527-539

10. Salmonson T, Danielson BG, Wikstrom B (1990) The Pharmacokinetics of Recombinant Human Erythropoietin after Intravenous and Subcutaneous Administration to Healthy Subjects. Br J clin Pharmac 29:709-713

Discussion - MEANINGFUL EVALUATION OF BIOTECHNOLOGY PRODUCTS

J.A. Galloway

I would like to ask what is the extent of communications between the regulatory agencies within the European Community and also to the FDA and what are the impediments to those communications?

P. Juul

As far as the EEC is concerned, we meet in Brussels at the Committee for Proprietary Medicinal Products eight times per year. We have six working parties which meet normally four to six times per year with representatives of each of the twelve member states discussing particular drugs or particular problems, producing guidelines or notices to applicants etc. With regard to communications with the FDA, I can mention that the European guidelines are sent to the FDA at the same time as they are sent to the industry and the national authorities for comments. So far, there have been one or two informal meetings between the two parties.

J.A. Galloway

What about any communication between the EEC and the Japanese regulatory authorities?

P. Juul

They also started. There have already been two meetings and a third one is planned.

R.G. Werner

Could you elaborate somewhat more about the objections of the EEC against mammalian cell cultures?

P. Juul

They have been raised by only two out of twelve countries. Denmark is one of them but we probably have to agree with the majority. Our concern about the use of mammalian cell lines, in this case for the production of human growth hormone, refers to the possible transferral of oncogenic DNA material to an injectable preparation that will be administered for many years to children. One should not forget that there are two other products at least on the European market produced by methods where we consider the risk of transfer of oncogenic material, including virusless. Of course, we accept mammalian cell lines; otherwise we would be without some of the more important drugs which can only be produced this way. The problem in this case is that we have a choice.

W.M. Wardell

I'd like to emphasize the subject of international harmonization, because I feel strongly that this is the only way to stop the drug development process from getting increasingly encumbered by the accretion of sometimes idiosyncratic national requirements. I have been to one or two meetings watching regulators talk about harmonization, but my growing feeling is that harmonization is not happening. I see regulators enthusiastically describing their own requirements but I do not see any pressure on them to harmonize.

P. Juul

One could almost say that so far, the EEC harmonization has meant that you add the twelve opinions, which means that the most severe and strict requirements are always the ones being accepted. On the other hand, one should not forget the agreement on the lack of necessity of LD50s, and I think that the European attitude about the longest duration of chronic animal experimentation, limiting it to six months may influence the FDA. At present it is very difficult to suggest changes, but personally I think that the usual type of carcinogenicity testing is in 99 out 100 cases a mere waste of resources, animals and time. This should probably be reviewed but it is difficult to resist asking for such a test since if the product turned out to be carcinogenic in man, we would be the ones to blame.

F. García Alonso

I would like to ask what is your opinion about the necessity to repeat clinical trials in different countries within the EEC, particularly in the case of biotechnology products such as human growth hormone or interferon.

P. Juul

I understand your concerns. Only exceptionally repetition in another EEC country is necessary. It does not make sense, for instance, to evaluate the effects of a new human growth hormone in patients with Turner's syndrome when they have been established with another product in another country. I think that an acute toxicity study and data from about ten patients switched from one of the preparations to another is all what we should ask.

L. Gauci

Shouldn't we focus more in postmarketing surveillance rather than in new clinical trials? Why hasn't this approach been followed more aggressively? Perhaps a common database of individuals followed several years should be considered.

P. Juul

I think that the main reason why postmarketing surveillance has not been approached more aggressively is that we do not have an exact idea of how to perform it. A database as you suggest could probably be envisaged in the case of some drugs, for instance recombinant human insulins. Hopefully the Pharmacovigilance Working Party shall come up with new ideas within the EEC.

PHARMACOLOGIC AND THERAPEUTIC ACTIONS OF BIOTECHNOLOGY PRODUCTS

© 1991 Elsevier Science Publishers B.V. (Biomedical Division)
The clinical pharmacology of biotechnology products.
M.M. Reidenberg, editor

BIOLOGICAL AND CLINICAL RESPONSE OF RECOMBINANT INTERFERON GAMMA IN PATIENTS WITH ADVANCED RENAL CELL CANCER

W. Aulitzky 1), Gastl G. 2), Aulitzky W.E. 2), Frick J. 2), Huber C. 1).

1) Department of Urology, General Hospital Salzburg, Austria

2) Division of Hematology, IIIrd Department of Internal Medicine, University Hospital Mainz

INTRODUCTION:

Interferon (IFN) gamma is a potent immunomodulatory agent with a wide range of biological properties (1). It enhances the functional activity of various immune effector cells. In addition, IFN gamma regulates the interaction of MHC restricted cytotoxic T cells with nonhemopoetic cells by inducing the synthesis of HLA products (2). Thus this compound might be an excellent candidate for modifying the hosts' immune response to malignant disease.

There is some evidence suggesting that immune mechanisms are critical for the control of renal cell cancer. First, spontaneous remissions are infrequently seen after tumor nephrectomy in patients with metastasizing RCC (3). Second, in almost all studies using biological response modifiers for treatment of advanced renal cell cancer tumor regressions were observed in some patients (4,5,6,13).

Most studies using biological response modifiers for treatment of advanced cancer applied the maximum tolerated dose assuming that this dose should be the optimum antiproliferative dose.

TABLE 1

n	Age	Sex	Karnovsky	TNM	Metastases	Response
1	75	m	70	T3N1M1	liver, local	MR
2	57	m	60	T4N2M1	liver, lung	PRO
3	66	m	100	T3NXM1	lung	SD
4	55	m	80	T4N4M1	lymph node,lung	PR
5	67	m	100	T4N1M1	lung	n.e.*)
6	68	m	90	T2NOM1	lung	SD
7	66	m	90	T4NOM1	lung	CR
8	57	m	90	T3N2M1	lung	PRO
9	57	m	100	T3NXM1	liver	PRO
10	50	f	90	T3NOM1	lung, CNS	PRO
11	69	m	70	T4NXM1	lung,lymph node	PRO
12	46	m	90	T4NXM1	lung	CR
13	41	m	70	T2NXM1	bone	n.e.*)
14	62	m	90	T3NOM1	lung,liver	PRO
15	54	m	90	T3NOM1	lung,bone	PRO
16	68	f	100	T3N2M1	lung	PR

Table 1. Patients characteristics and clinical response after 12 months.

However, there is increasing evidence from experimental animal studies and clinical trials that the maximum tolerable dose does not represent the optimum immunomodulatory dose (7,8,9). Thus, if immunomodulation is one of the critical mechanisms of action for antitumor activity, new strategies are required for identification and definition of biologically and clinically effective doses of cytokines for cancer treatment.

MATERIAL AND METHODS:

The study reported was performed according to the Declaration
of Helsinki. The protocols were approved by the local Ethics
committee. All patients had given their written informed consent
prior to start of treatment.

16 patients suffering from progressive metastasizing renal cell
carcinoma were treated with 10, 100 and 500 µg recombinant
interferon gamma. Fourteen of the patients were male and 2 female.
Their ages ranged from 41 to 75 years (median 59). Major mani-
festation of the disease were lung, bone and liver metastases.
The clinical characteristics of the patients are summarized
in table 1 (Pat 1-16).

Recombinant IFN gamma, originally produced by Genentech Inc.,
with a specific activity of 2x10 E7 I.U./mg protein was obtained
from Boehringer Ingelheim International (Ingelheim, FRG). Some
of the data were reported in detail previously (13,14,15).

TREATMENT PROTOCOL

Treatment consisted of two consecutive phases. During an initial
dose finding phase three different doses of r-IFN gamma
(10, 100, 500 µg) were administered subcutaneously (s.c.). Each dose
was administered 3-times to the same individual. The sequence
of the dose levels was randomly assigned. In order to prevent
carry over effects a therapy -free interval of 2 weeks was
introduced between each dose level.

Blood was drawn 1 hour before and 4, 12, 24, 48, 72, 96, 120
and 167 hours after the IFN-gamma injections.

EVALUATION OF TUMOR RESPONSE AND DRUG-INDUCED SIDE EFFECTS:

Tumor response and severity of side effects were defined according to WHO criteria (10). Quantitative changes of tumor size
were assessed at least every four weeks by clinical examination,
abdominal sonography, chest x-ray studies and CT scanning. All
patients were assessed by more than one method of evaluation.

LABORATORY INVESTIGATIONS:

Differential blood counts were performed using established
standard methods. Briefly, cell counts were assessed with a
coulter counter and differential blood counts were performed
on Giemsa stained blood smears by counting a minimum of 200
cells.

Numbers of leukocyte subsets in the periphery were determined
by staining cytocentrifuge preparations of Ficoll separated
peripheral blood mononuclear cells with an indirect immune-
peroxidase method using Leu3, Leu2a, Leu1, Leu7.
Anti HLA DR (Beckton Dickinson Inc., USA) as primary antibodies.

Serum levels of TNF alpha were determined by commercially
available immunoradiometric assays (IRMA) (Medgenix Brusseles,
Belgium). The lower limit of detection using these tests is
$>$ 5 pg/ml for TNF alpha and $>$ 0.2 U/ml (NIH standard) for IFN
gamma. Neopterin serum levels were determined by a commercially
available radioimmunoassay (Henning Inc., Berlin FRG).

STATISTICS:

Statistical significance of the differences before and after

application of the cytokines were analyzed by means of the
confidence limits of the mean difference (13) or by Wilcoxon
matched-pairs signed-rank test and Mann-Whitney's U test using
a commercially available statistical software package
(SPSS Inc, Chicago, USA).

RESULTS:

CLINICAL RESPONSE:

Two of the 16 patients enrolled did not complete the dose finding
phase and were therefore excluded from response evaluation:
patient no. 5 experienced rapid disease progression during the
first three weeks of therapy and patient no. 13 refused further
treatment after six weeks due to severe constitutional symptoms
following administration of 500 μg of IFN-gamma.
14 patients completed the dose finding phase and entered the
phase II efficacy trial receiving r-IFN-gamma 100 μg s.c. once
weekly. During a median time of treatment of ten months
(range, 2-32 months) complete or partial responses were observed in 4 of 14 evaluable patients. In all 4 responses were
seen within three months of initiating IFN-gamma. All 4 complete
and partial responses were seen in patients with lung metastases (table 1).

SIDE EFFECTS:

During the dose-finding phase fever, fatigue and chills were the
most frequent side effects. Both frequency and severity of these
symptoms were clearly dose-dependent. Grade 2 toxicity (WHO-
grading) was observed in twelve of 14 cases during treatment

194

STUDY DESIGN

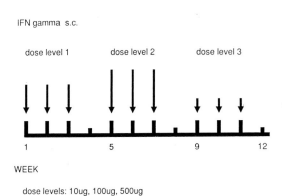

IFN gamma s.c.

dose level 1   dose level 2   dose level 3

1   5   9   12

WEEK

dose levels: 10ug, 100ug, 500ug

TNF

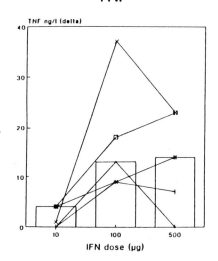

Figure 1. Study design

Figure 2. Dose depending of the increase of
TNF release after treatment with IFN-γ. Delta
values are calculated from pretreatment and peak values. Bars represent
median values.

with 500 µg of IFN-gamma. This was seen in half of the patients

receiving 100 µg of IFN-gamma and in only three patients at

the 10 µg dose level.

Two patients refused further treatment due to side effects after

500 µg r-IFN-gamma.

**TUMOR NECROSIS FACTOR ALPHA SERUM LEVELS AFTER TREATMENT WITH
RECOMBINANT INTERFERON GAMMA:**

TNF alpha serum levels were measured in 282 serum samples of 5

patients. Only two out of five patients had detectable serum levels

of TNF alpha in pretreatment samples. Administration of recombinant

IFN gamma led to a statistically significant increase of TNF

alpha serum levels in almost all cases (p 0.01) (Fig. 2).

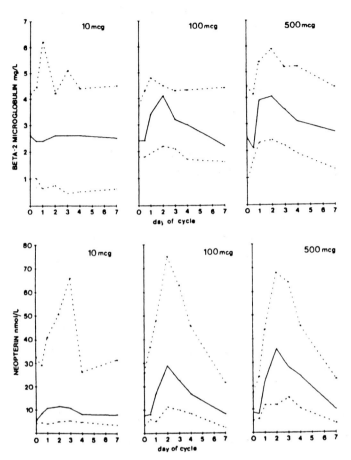

Figure 3. Neioterin
and Beta-2-Micro-
globulin release
after single
injections of IFN-
gamma in 16 patients
with m-RCC (median
values and range).

TNF levels peaked 24 hours after application of r-IFN gamma
and returned to baseline values within four days.

NEOPTERIN AND BETA-2-MICROGLOBULIN AFTER TREATMENT WITH
r-IFN GAMMA:

Neopterin is a metabolic product of monocytes activated by IFN
gamma (11). Neopterin levels increased in all patients after

TABLE 2

IFN-γ dose	10 µg	100 µg	500 µg
Leukocytes			
4 h	101% (74-129)	82% (64-118)[a]	98% (46-123)
24 h	88% (55-113)	83% (45-149)[b]	73% (32-97)[b]
Granulocytes			
4 h	102% (66-165)	99% (66-146)	97% (48-201)[c]
24 h	101% (60-140)	90% (41-187)[a]	76% (30-131)[b,c]
Lymphocytes			
4 h	96% (36-195)	85% (54-145)	69% (29-196)[b]
24 h	75% (43-137)[a]	66% (23-114)[b]	65% (39-145)[b,d]
Monocytes			
4 h	90% (7-405)	57% (0-215)[a]	35% (0-400)[b,e]
24 h	127% (0-392)	66% (0-459)	54% (0-357)[b]

[a] $p < 0.05$ when compared to pretreatment values.

[b] $p < 0.01$ when compared to pretreatment values.

[c] $p < 0.01$ when response to 10 µg and to 500 µg are compared.

[d] Not significant when response to 10 µg and to 500 µg are compared.

[e] $p = 0.055$ when response to 10 µg and to 500 µg are compared.

Table 2. Response of white blood cell counts to single doses of IFN-γ expressed as a percentage of pretreatment values.

100 µg and 500 µg of IFN gamma peaking at 48 hours after the application (Fig 3). As shown in the figure, nearly maximum

TABLE 3

IFN-γ dose	10 µg	100 µg	500 µg
Leu1 T lymphocytes			
4 h	109% (50-177)	116% (43-173)[a]	84% (30-167)[a]
24 h	88% (46-151)	66% (14-151)	79% (23-157)[b]
Leu3 T lymphocytes			
4 h	132% (50-238)	133% (53-215)	90% (40-283)
24 h	86% (46-177)	75% (20-154)	82% (26-154)
Leu2a T lymphocytes			
4 h	88% (38-180)	122% (60-159)	76% (25-114)[c,a]
24 h	74% (32-185)	61% (13-183)[c]	63% (23-193)[c,a]
Leu7* NK cells			
4 h	82% (32-220)[c]	63% (24-197)[c]	38% (0-139)[c,a]
24 h	69% (18-414)[c]	39% (8-351)[b]	32% (6-130)[b,a]
HLA DR* cells			
4 h	70% (20-212)	51% (21-245)[c]	38% (25-127)[b,a]
24 h	88% (56-132)	85% (35-249)	70% (39-176)[c]
T4/T8 ratio			
4 h	130% (44-271)	118% (76-205)	126% (66-217)[c]
24 h	121% (46-234)	110% (77-264)	122% (56-326)[b]

[a] $p < 0.05$ when response to 10 µg and 500 µg are compared.

[b] $p < 0.01$ when compared to pretreatment values.

[c] $p < 0.05$ when compared to pretreatment values.

Table 3. Response of T-lymphocytes subsets to single doses of r-IFN-γ expressed as a percentage of pretreatment values.

induction was achieved with 100 µg r-IFN-gamma and increasing
the dose to 500 µg r-IFN-gamma caused only a marginally higher
induction of the synthesis of this molecule. An identical pattern
was observed with beta-2-microglobulin after treatment with single
doses of r-IFN-gamma (16). Significant induction was observed
already after treatment with 100 µg r-IFN gamma. Administration
of 500 µg r-IFN-gamma was followed by a slightly higher peek.

ACUTE EFFECTS ON THE COMPOSITION OF PERIPHERAL BLOOD LEUCOCYTES:

IFN-gamma treatment caused only minor effects on total white
blood cell counts. However the relative composition of various
leucocyte subsets changed markedly upon treatment with IFN-gamma.
Granulocyte counts were not effected, whereas the numbers of
lymphocytes and monocytes decreased after administration of
cytokines reaching the nadir two to twentyfour hours after injection.
The highest dose of IFN-gamma lowered circulating monocytes by
approximately 70 % and circulating lymphocytes marginally ($\sim$ 30 %).
Numbers returned to pretreatment values within 48 hours.

T-CELL SUBSETS:

Lymphocyte subsets after IFN-γ administration showed only minor
changes. The number of circulation Leu 1* cells declined 24 hours
after 500 µg IFN-gamma from a median value of 629 cells/µl to
608 (p$<$ 0.01) (Table 3). The reduction after 10 µg and 100 µg did
not reach the level of statistical significance (Table 2).
Leu3* cells increased slightly 4 h after application of the
two lower dose levels, whereas 500 µg resulted in decreasing cell
counts. In contrast, the number of Leu2a* cells in the peripheral

blood fell significantly with a nadir 24 h after administration
of both 100 µg and 500 µg of IFN-γ (from a median of 220
cells/µl to 120 cells/µl, p $<$ 0.05). The divergent behavior of
Leu3* and Leu2a* cells resulted in elevated T4/T8 ratios.

DISCUSSION:

Three major conclusions can be derived from the results of this
study: First, significant modification of biological response
was observed at dose levels far below the maximum tolerated
dose. r-IFN gamma effects in this low dose range were demonstrated on homing properties of peripheral blood leucocytes,
synthesis of HLA proteins and neopterin, and activation of the
cytokine network. These findings are in conformity with results
of other groups, were an optimum immunomodulatory dose of 100 µg
was proposed (9). Second treatment of patients with advanced renal
cell cancer with low dose of r-IFN gamma can induce tumor
regressions in these patients. Moreover, the remission rate
observed in our study was within the range as those observed
after treatment of RCC patients with IFN alpha, IL-2, IL-2 plus
IFN alpha, IL-2 + LAK cells (4,5,6). However, patients treated
with low dose gamma interferon experienced considerably less
toxicity than reported from studies applying tolerable doses of
cytokines. Previous reports of r-IFN gamma for treatment of
advanced RCC demonstrated an extremely variable antitumor
activity in RCC patients with response rates ranging from 0 to
30 % (12). It cannot be excluded that differences in patient
selection account for varying clinical results. Nevertheless,
the fact that responses were seen almost exclusively in patients
treated intermittently with r-IFN gamma might indirectly support
the view, that continuous intravenous treatment with maximum

tolerable doses is not the appropriate way for using r-IFN gamma in these patients. We therefore conclude that the low dose range of r-IFN gamma has to be further explored in these patients.

The last important observation of this study is that biological response seems to be a prerequisite for tumor response in RCC patients. All patients who showed no signigicant increase of beta-2-microglobulin levels after treatment with r-IFN gamma were resistant to therapy (13). Thus these marker molecules might be appropriate parameters for the action of r-IFN gamma and be thereby a means of designing optimal immunomodulatory treatment schedules.

REFERENCES

1. Gastl G, Huber C (1988). The Biology of Interferon Action. Blut 56:193-199.

2. Niederwieser D., Auböck J., Troppmair J., Beck P., Fritsch P., Huber Ch. (1988). IFN mediated induction of MHC expression on human keratinocytes and its influence on in vitro immune response. J Immunol, 140:2556-2564

3. Oliver RTD, Miller RM, Mehta A, Barnett MJ. A Phase II study of surveillance in patients with metastatic renal cell carcinoma and assessment of response of such patients to therapy on progression. Mol Biother 1988 1:14-20

4. Rosenberg SA, Lotze MT, Muul LM et al (1988). A progress report on the treatment of 157 patients with advanced cancer using lymphokine activated killer cells and interleukin 2 or high dose interleukin 2 alone. N Eng J Med 316:889-897

5. Neidhart JA (1986) Interferon therapy for the treatment of renal cell cancer. Cancer 57:1696-1699

6. Quesada JR. (1988) Biological response modifiers in the therapy of metastatic renal cell carcinoma. Seminars in Oncology 15:396-407

7. Talmadge JE, Tribble HR, Pennington RW, Philipps H, Wiltrout RH. (1987): Immunomodulatory and immunotherapeutic properties of recombinant gamma interferon and recombinant tumor recrosis factor in mice. Cancer Res, 47:2563-2570

8. Kleinerman ES, Kurzrock R., Wyatt D, et al. Activation or suppression of the tumoricidal properties of monocytes from cancer patients following treatment with human recombinant interferon gamma. Cancer Res 46:5401-5405

9. Maluish A., Urba W.J., Longo D.L. et al. (1988). The determination of an immunologically active dose of interferon gamma in patients with melanoma. J Clin Onc 6:434-445

10. WHO handbook for reporting results of cancer treatment. WHO, Geneva, 1979

11. Huber C, Batchelor SR, Fuchs D, Hausen A, Lang A, Niederwieser D, Reibnegger G, Swetly P, Troppmair J, Wachter H (1984): Immune response associated production of neopterin. Release from macrophages primarily under control of IFN gamma. J Exp Med, 160:310-316

12. Aulitzky W.E., Aulitzky W., Lttichau I., Huber CH. Treatment of advanced renal cell carcinoma with recombinant Interferon gamma. Interferons and cytokines (in press).

13. Aulitzky W., Gastl G., Aulitzky W.E., Herold M., Kemmler J., Mull B., Frick J. and Huber C. (1989). Successfull treatment of metastatic renal cell carcinoma with a biologically active dose of recombinant interferon gamma. J Clin Oncol,7, 1875-1884.

14. Aulitzky W.E., Aulitzky W., Gastl G., Lanske B., Reitter J., Frick J., Tilg H., Berger M., Herold M. and Huber C. (1989). Acute effects of single doses of recombinant IFN gamma on blood cell counts and lymphocyte subsets in patients with advanced renal cell cancer. J.Interferon Res. 9, 425-433.

15. Aulitzky W.E., Aulitzky W., Frick J., Herold M., Gastl G.,
    Tilg H., Berger M. and Huber C. (1988). Treatment of cancer
    patients with recombinant interferon gamma induces release of
    endogenous tumor necrosis factor alpha. Immunobiology 1990
    180:385-394

16. Cytokines in the control of beta-2-microglobulin release 2. In vi⁣
    studies with recombinant interferons and antigens K. Nachbaur,
    J. Troppmair, B. Kotlan, P. König, W. Aulitzky, P. Bieling,
    Ch. Huber Immunbiol. 1988, Vol. 177, S 66-75

Discussion -BIOLOGICAL AND CLINICAL RESPONSE OF RECOMBINANT INTERFERON GAMMA IN PATIENTS WITH ADVANCED RENAL CELL CANCER

M.M. Reidenberg

I think this was a very important paper. The traditional cytotoxic drugs used in cancer have an ordinary dose response relationship and it was really Dr. Skipper popularizing the fractional kill hypothesis that lead to the present conventional treatment of cancer using the highest tolerated dose of multiple cytotoxic drugs. When cytokines were then introduced into therapeutics the same concept of really pushing the dose to toxicity and then backing off a little bit was used. And the results were disappointing. Now, Dr. Aulitzky is presenting really good clinical pharmacological data suggesting that the conceptual basis of treating cancer based on the normal dose response of cytotoxic drugs and the Skipper Hypothesis just isn't appropriate when we are dealing with biological response modifiers.

L. Gauci

I remember the arguments that occurred in two of the major pharmaceutical companies that were involved firstly with IFN alpha and then with IFN gamma development. There was one opinion which said there was only one way to develop a cancer drug and that is to give it at high doses like chemotherapy, but there was a small group of timorous people who said that these were biologicals and perhaps we ought to target them differently. The development of alpha IFN was rescued because of a rare selective indication, hairy cell leukemia, and gamma IFN was saved because of congenital granulomatosis.

W. Aulitzky

I think, that we have to make an important step in clinical research. We have to go back into man as a big test tube. We have to do it very carefully and we must not do any harm. we also have to go back into man because that is the only environment where we can study all those effects and interactions we need to study. We can't do it in simple in vitro systems.

© 1991 Elsevier Science Publishers B.V. (Biomedical Division)
The clinical pharmacology of biotechnology products.
M.M. Reidenberg, editor

TISSUE-TYPE PLASMINOGEN ACTIVATOR

H. ROGER LIJNEN AND DESIRE COLLEN

Center for Thrombosis and Vascular Research, Campus Gasthuisberg, O & N, Here-
straat 49, B-3000 Leuven (Belgium)

INTRODUCTION

Thrombotic complications of cardiovascular diseases, such as acute myocardial infarction, cerebrovascular thrombosis and venous thromboembolism, constitute a main cause of death and disability. Thrombolytic therapy, which consists in the administration of plasminogen activators which activate the fibrinolytic system in blood, could favorably influence the outcome of these life-threatening diseases. The fibrinolytic system contains a proenzyme, plasminogen, which by the action of plasminogen activators is converted to the active enzyme plasmin, which in turn digests fibrin to soluble degradation products. Inhibition of the fibrinolytic system occurs both at the level of the plasminogen activators, by plasminogen activator inhibitors (mainly PAI-1 and PAI-2) and at the level of plasmin, mainly by α_2-antiplasmin.

Currently, five thrombolytic agents are either approved for clinical use or under clinical investigation in patients with acute myocardial infarction. These include streptokinase, two chain urokinase (tcu-PA), anisoylated plasminogen streptokinase activator complex (APSAC), tissue-type plasminogen activator (t-PA) and single chain urokinase-type plasminogen activator (scu-PA, prourokinase). Streptokinase, APSAC and two chain urokinase cause extensive systemic activation of the fibrinolytic system, which may result in degradation of several plasma proteins, including fibrinogen, Factor V and Factor VIII.

Physiological fibrinolysis, however, is highly fibrin-specific as a result of molecular interactions between the components of the fibrinolytic system (1). The physiological plasminogen activators, t-PA and scu-PA, activate plasminogen preferentially at the fibrin surface. Plasmin, associated with the fibrin surface, is protected from rapid inhibition by α_2-antiplasmin and may thus efficiently degrade the fibrin of a thrombus (1,2). This fibrin-specific mechanism of action of t-PA has triggered great interest in the use of this agent for therapeutic thrombolysis. Production of t-PA by recombinant DNA technology (rt-PA) (3,4) has made it available for large scale clinical use. This chapter will review the biochemical and biological properties of rt-PA and its use for thrombolysis.

BIOCHEMICAL PROPERTIES OF TISSUE-TYPE PLASMINOGEN ACTIVATOR (t-PA)

Sources of t-PA

The first satisfactory purification of human t-PA was obtained from uterine tissue (5). Using an antiserum raised against uterine plasminogen activator, it was shown that tissue plasminogen activator, vascular plasminogen activator, and blood plasminogen activator are immunologically identical, but different from urokinase (6). The plasminogen activator found in blood is identical to vascular plasminogen activator, which is synthesized and secreted by endothelial cells (7) and is now generally called "tissue-type plasminogen activator" (t-PA). t-PA has been purified from the culture fluid of a stable human melanoma cell line (Bowes, RPMI-7272) in sufficient amounts to permit the study of its biochemical and biological properties (8).

The cDNA of human t-PA has been cloned and expressed in E. coli and in mammalian cell systems (3,9,10). The generation of Chinese hamster ovary cells capable of producing human t-PA has allowed the development of large-scale tissue culture fermentation and purification procedures (11). The resulting product (Activase[R]) consists primarily of a single chain molecule. t-PA for clinical use is presently produced by recombinant DNA technology (Activase[R], Genentech Inc. or Actilyse[R], Boehringer Ingelheim GmbH).

Structural properties of t-PA

t-PA is a serine proteinase with a molecular weight of about 70,000, consisting of a single polypeptide chain of 527 amino acids with Ser as the NH_2-terminal amino acid, as deduced from the cDNA sequence (3). Plasmin converts t-PA to a two chain molecule by hydrolysis of the Arg^{275}-Ile^{276} peptide bond. The NH_2-terminal region (heavy chain) is composed of multiple structural-functional domains, including a "finger-like" domain (F) homologous to the finger domains in fibronectin, an "epidermal growth factor domain" (E) homologous to human epidermal growth factor, and two disulphide bonded triple loop structures commonly called "kringles" (K_1 and K_2), homologous to the kringle regions in plasminogen (3). The COOH-terminal region (light chain), comprising residues 276 to 527, is homologous to other serine proteinases and contains the catalytic site, which is composed of His^{322}, Asp^{371} and Ser^{478} (3). The structures required for the enzymatic activity of t-PA are fully comprised within the COOH-terminal polypeptide chain, as evidenced by the intact activity of the isolated chain, separated chemically (12,13) or prepared by recombinant DNA technology (14,15).

These structural domains of t-PA are involved in most of its functions and interactions, including its enzymatic activity, binding to fibrin, stimulation of plasminogen activation by fibrin, binding to receptors, and inhibition by plasminogen activator inhibitors.

Functional properties of t-PA

The structures involved in the fibrin-binding of t-PA are fully comprised within the NH_2-terminal (heavy) chain, as evidenced by the intact fibrin-affinity of the heavy chain isolated after mild reduction of two chain t-PA (12,13). Evidence obtained with domain deletion mutants of t-PA indicated that its affinity for fibrin is mediated via the finger domain and mainly via the second kringle domain (16). A lysine-binding site is involved in the interaction of the kringle-2 domain but not of the finger domain with fibrin (14). Gething, et al. (17) have however suggested that the kringle-1 and kringle-2 domains of t-PA would be equivalent in their affinity for fibrin, although the kringle-1 domain does not contain a lysine-binding site. The presence of a weaker lysine-binding site in kringle 2 similar to the "AH-site" (aminohexyl-site) in plasminogen has also been suggested (18). This AH-site would interact with internal lysine residues in the fibrin matrix, whereas the lysine-binding site would interact with COOH-terminal lysine residues. In the process of fibrinolysis, binding of t-PA to intact fibrin may initially be mediated by the F domain, while upon partial fibrin digestion by plasmin, increased binding of t-PA to newly exposed COOH-terminal lysine residues may occur via the lysine-binding site in the K_2 domain (16). Because of its AH site, the K_2 domain may also play a role in the initial binding to intact fibrin.

Mechanism of action of t-PA

t-PA is a poor enzyme in the absence of fibrin, but the presence of fibrin strikingly enhances the activation rate of plasminogen (19). The kinetic data support a mechanism in which fibrin provides a surface to which t-PA and plasminogen adsorb in a sequential and ordered way yielding a cyclic ternary complex. Fibrin essentially increases the local plasminogen concentration by creating an additional interaction between t-PA and its substrate. The high affinity of t-PA for plasminogen (low K_m) in the presence of fibrin thus allows efficient activation on the fibrin clot, while plasminogen activation by t-PA in plasma is a comparatively inefficient process (19). However, others have claimed that fibrin influences both the K_m and k_{cat} of the activation of plasminogen by t-PA (20).

Plasmin formed on the fibrin surface has both its lysine-binding sites and active site occupied and is thus only slowly inactivated by α_2-antiplasmin (half-life of about 10-100 s); in contrast, free plasmin, when formed, is rapidly inhibited by α_2-antiplasmin (half-life of about 0.1 s) (2). The fibrinolytic process thus seems to be triggered by and confined to fibrin.

THROMBOLYTIC PROPERTIES OF t-PA

In 1981, the first patients were treated with t-PA obtained from the Bowes melanoma cell line. Intravenous administration of 7.5 mg t-PA over 24 hours induced complete lysis of a renal and iliofemoral thrombosis in a renal trans-plant patient without systemic fibrinolytic activation or bleeding (21). How-ever, intravenous infusion of 5 to 25 mg of t-PA over 24 to 36 hours did not produce thrombolysis in four patients with deep vein thrombosis (22). A pilot-study of melanoma cell t-PA was carried out in 7 patients with acute myocardial infarction. Coronary thrombolysis in the absence of fibrinogen breakdown was achieved in 6 of these 7 patients (23).

These encouraging initial results have triggered the organization of clinical trials using recombinant t-PA in several indications, including deep vein throm-bosis, major pulmonary embolism, arterial thromboembolism and acute thromboem-bolic stroke. Most experience has, however, been obtained in the treatment of coronary artery disease. Reduction of infarct size, preservation of ventricular function and/or decreased mortality have been demonstrated in patients with acute myocardial infarction following administration of streptokinase, APSAC and rt-PA (for references, cfr. 24,25). This beneficial effect of thrombolysis will probably also hold for other plasminogen activators.

Coronary reperfusion is very likely the most important, if not the only, sig-nificant contributor to preservation of ventricular function and to reduction in mortality, although the magnitude of the effect appears to be time dependent. Indeed, subgroup analyses of patients with and without successful reperfusion with streptokinase or rt-PA have consistently demonstrated a lower mortality rate in successfully reperfused patients, and a better outcome if reperfusion is obtained early. The most rational treatment of patients with acute myocardial infarction is therefore likely to be thrombolytic therapy with agents that reperfuse the most coronary arteries as rapidly as possible.

The 2 comparative trials to date which have examined the reperfusion effica-cies of rt-PA and streptokinase have been evaluated by meta-analysis (26). These results show that rt-PA appears more effective than streptokinase for the early recanalisation of occluded coronary arteries, both when given within 3 hours of the onset of symptoms and when given later. Comparable effects on the preservat-ion of left ventricular function have been demonstrated for rt-PA and streptoki-nase in 2 comparative trials, but futher studies would assist in determining whether differences in early coronary reperfusion rates may translate into a comparably better outcome. Much larger, direct comparative studies are needed before scientifically valid statements can be made regarding the relative effi-cacy of the available thrombolytic agents for the reduction of both moribidity and mortality. Such studies, i.e. GISSI-II are presently carried out.

ADMINISTRATION SCHEMES OF rt-PA

Due to its short in vivo half-life (initial $t_{\frac{1}{2}}$ of 4 min and terminal $t_{\frac{1}{2}}$ of 46 min in patients with acute myocardial infarction (27)) relatively large amounts of rt-PA have been administered as a continuous intravenous infusion. In clinical trials in patients with acute myocardial infarction, total doses of rt-PA ranging between 40 and 150 mg, have been administered. The currently used or recommended dose is 100 mg, to be given intravenously as a 10 mg bolus in 1 to 2 minutes followed by 50 mg in the first hour, 20 mg over the second hour and 20 mg over the third hour. For patients weighing less than 65 kg, a dose of 1.25 mg per kg, administered over 3 hours as indicated above, may be used. A dose of 150 mg has been associated with an increased frequency of intracranial bleeding and should not be used.

Neuhaus et al. (28) have administered an accelerated dosage regimen of rt-PA (15 mg intravenous bolus, 50 mg infusion over 30 min and 35 mg infusion over the following 60 min) to patients with acute myocardial infarction of less than 6 hours duration. Patent infarct-related arteries were observed in 74% of the patients at 60 min after the start of the infusion, in 91% at 90 min and in 92.4% at 24 hours. Rapid infusion of 100 mg rt-PA over 90 min thus yielded a high early patency rate, and was not associated with an increase in reocclusion or adverse effects (28).

Bolus administration of rt-PA has also been performed in patients with acute myocardial infarction. Tebbe et al. (29) have given a single 50 mg bolus of rt-PA over 2 min to 20 patients and obtained patency in 75% of those patients. Reocclusion occurred in 22% of patients and 1 patient died due to intracranial hemorrhage. rt-PA plasma levels increased to 9.8 ± 3.6 μg/ml and the bolus caused a decrease of circulating fibrinogen to 55% of baseline after 2 to 4 hours. Tranchesi et al. (30) have administered rt-PA as a bolus of 50 mg, 60 mg or 70 mg and concluded that a bolus of 70 mg yielded results comparable to the conventional infusion scheme (72% recanalization at 60 min), and was associated with only minor bleeding complications. These preliminary results in patients with acute myocardial infarction suggest that bolus administration of t-PA may yield similar patency rates as the infusion regimen.

Combined administration of t-PA and scu-PA has been performed in patients with acute myocardial infarction in order to investigate if the effect on clot dissolution might be more than additive. Synergism may allow a reduction of the total dose of thrombolytic agent and may reduce the occurrence of haemostatic side effects. Preliminary results in a small number of patients (31,32) have suggested that combining t-PA and scu-PA or tcu-PA at approximately 20% of their individual therapeutic dose, produced coronary artery reperfusion without associated fibrinogenolysis. Such synergistic effect of t-PA and tcu-PA on coronary reperfusion has, however, not been confirmed in a larger study, although the use of the combination was associated with a reduction of the frequency of reocclu-

210

sion (33). A subsequent study in patients with acute myocardial infarction, using 20 mg rt-PA combined with 10, 15 or 20 mg rscu-PA given intravenously over 90 min, yielded recanalization at 90 min in only 31 to 41% of patients (34). Combined administration at reduced doses thus was less efficient than that of either drug at the presently recommended dose.

REFERENCES

1. Collen D (1980) Thromb Haemost 43:77-89

2. Wiman B, Collen D (1978) Nature 272:549-550

3. Pennica D, Holmes WE, Kohr WJ, Harkins RN, Vehar GA, Ward CA, Bennett WF, Yelverton E, Seeburg PH, Heyneker HL, Goeddel DV, Collen D (1983) Nature 301:214-221

4. Vehar GA, Spellman MW, Keyt BA, Ferguson CK, Keck RG, Chloupek RC, Harris R, Bennett WF, Builder SE, Hancock WS (1986) Cold Spring Harbor Symposia on Quantitative Biology LI:551-562

5. Rijken DC, Wijngaards G, Zaal-De Jong M, Welbergen J (1979) Biochim Biophys Acta 580:140-153.

6. Rijken DC, Wijngaards G, Welbergen J (1980) Thromb Res 18:815-830

7. Levin EG, Loskutoff DJ (1982) J Cell Biol 94:631-636

8. Rijken DC, Collen D (1981) J Biol Chem 256:7035-7041

9. Kaufman RJ, Wasley LC, Spiliotes AJ, Gossels SD, Latt SA, Larsen GR, Kay RM (1985) Molecular and Cellular Biology 5: 1750-1759

10. Sambrook J, Hanahan D, Rodgers L, Gething MJ (1986) Mol Biol Med 3:459-481

11. Builder SE, Grossbard E (1986) In: Murawski, Peetoom (eds), Transfusion Medicine: Recent Technological Advances, AR Liss, New York, pp 303

12. Holvoet P, Lijnen HR, Collen D (1986) Eur J Biochem 158:173-177

13. Rijken DC, Groeneveld E (1986) J Biol Chem 261:3098-3102

14. van Zonneveld AJ, Veerman H, Pannekoek H (1986) Proc Natl Acad Sci USA 83:4670-4674

15. Verheijen JH, Caspers MPM, Chang GTG, De Munk GAW, Pouwels PH, Enger-Valk BE (1986) EMBO J 5:3525-3530

16. van Zonneveld AJ, Veerman H, Pannekoek H (1986) J Biol Chem 261:14214-14218

17. Gething MJ, Adler B, Boose JA, Gerard RD, Madison EL, McGookey D, Meidell RS, Roman LM, Sambrook J (1988) EMBO J 7:2731-2740

18. Verheijen JH, Caspers MPM, de Munk GAW, Enger-Valk BE, Chang GTG, Pouwels PH (1987) Thromb Haemost 58:491 (Abstract 1814)

19. Hoylaerts M, Rijken DC, Lijnen HR, Collen D (1982) J Biol Chem 257:2912-2919

20. Nieuwenhuizen W, Voskuilen M, Vermond A, Hoegee-de Nobel B, Traas DW (1988) Eur J Biochem 174:163-169

21. Weimar W, Stibbe J, Van Seyen AJ, Billiau A, De Somer P, Collen D (1981) Lancet ii:1018-1020

22. Verstraete M, Collen D (1985) In: Collen D, Lijnen HR, Verstraete M (eds), Thrombolysis: Biological and Therapeutical Properties of New Thrombolytic Agents. Contemporary Issues in Haemostasis and Thrombosis. Churchill Livingstone, Edinburgh, pp 49-60

23. Van de Werf F, Ludbrook PhA, Bergmann SR, Tiefenbrunn AJ, Fox KAA, De Geest H, Verstraete M, Collen D, Sobel B (1984) N Engl J Med 310:609-613.

24. Collen D, Lijnen HR, Todd PA, Goa KL (1989) Drugs 38:346-388

25. Collen D, Lijnen HR (1990) Biochem Pharmacol 40:177-186

26. Chesebro JH, Knatterud G, Braunwald E (1988) N Engl J Med 319:1544-1545

27. Garabedian HD, Gold HK, Leinbach RC, Johns JA, Yasuda T, Kanke M, Collen D (1987) J Am Coll Cardiol 9:599-607

28. Neuhaus K-L, Feuerer W, Jeep-Tebbe S, Niederer W, Vogt A, Tebbe U (1989) J Am Coll Cardiol 14: 1566-1569

29. Tebbe U, Tanswell P, Seifried E, Feuerer W, Scholz K-H, Herrmann KS (1989) Am J Cardiol 64:448-453

30. Tranchesi B, Verstraete M, Vanhove Ph, Van de Werf F, Chamone DF, Bellotti G, Pileggi F (1990) Coronary Artery Disease 1:83-88

31. Collen D, Stump DC, Van de Werf F (1986) Am Heart J 112:1083-1084

32. Collen D, Van de Werf F (1987) Am J Cardiol 60:431-434

33. Topol EJ, Califf RM, George BS, Kereiakes DJ, Rothbaum D, Candela RJ, Abbotsmith CW, Pinkerton CA, Stump DC, Collen D, Lee KL, Pitt B, Kline EM, Boswick JM, O'Neill WW, Stark RS (1988) Circulation 77: 1100-1107

34. Tranchesi B, Bellotti G, Chamone DF, Verstraete M (1989) Am J Cardiol 64:229-232

Discussion - TISSUE-TYPE PLASMINOGEN ACTIVATOR

P. Juul

Can you comment on the relationship between time of administration of t-PA and clinical response?

H.R. Lijnen

Already some of the earlier clinical trials suggested that it is very important to give the thrombolytic agent as soon as possible after the onset of myocardial infarction. It was then suggested that six hours after the onset of acute myocardial infarction t-PA would not be useful anymore to recanalize an occluded coronary artery. Several studies now underway are evaluating the effect of late recanalization of coronary arteries on clinical outcome.

M.M. Reidenberg

What is the difference between the thrombus that readily lyses and the ones that apparently do not even where the t-PA is given shortly after the clinical event?

H.R. Lijnen

Well, one of the possibilities is that a thrombus that is very rich in platelets would be much more resistant to thrombolysis. It is also known that an older thrombus is much more resistant to lysis than a younger thrombus, probably because of a higher extent of organization.

M.M. Reidenberg

But does organization occur within the first few hours?

H.R. Lijnen

Probably not.

M.M. Reidenberg

But yet this is the time period we are talking about. One cannot claim that a 30% failure in recanalization is due to the organization of the thrombus.

H.R. Lijnen

No. That's probably because thrombi which are very rich in platelets are very resistant to lysis. Actually it may be that recanalization in 100% of patients is an impossible goal. There may always be 20% or 25% of patients that we will not be able to recanalize because of the resistance of the thrombus.

J. Bigorra

Do you have separate data for patients older than 60 or 65? What was the upper limit in this trial?

H.R. Lijnen

In the first clinical trials with t-PA the age limit was usually 65 to 75 years. In the GISSI II trial, however, more than 20 percent of the patients were over 70. The mortality in patients over 70 was much higher, but was not different between t-PA and streptokinase.

H.J. Röthig

Do you think the decrease in the levels in fibrinogen is bad in general? This reduces viscosity and most of the cardiologists are very keen on finding drugs which are doing this and probably this might be one of the advantages of streptokinase therapy?

H.R. Lijnen

A lot of people feel that to spare fibrinogen in a thrombolytic therapy is not really an advantage. And that there is no real correlation between frequency of bleeding complications and reduction of fibrinogen.

P. Tanswell

I think that the supposition that the use of streptokinase reduces blood viscosity is not based on fact. Although streptokinase does completely degrade fibrinogen this does not have a significant effect on blood viscosity. And the other point to be made regarding fibrinogen specificity is that this probably is responsible for the higher efficacy of t-PA even if a significantly lower bleeding tendency has not yet been fully demonstrated.

H.R. Lijnen

There is a difference in the nature of the fibrinogen degradation products that one gets with t-PA and streptokinase. With t-PA one gets much more early degradation products, the large X and Y fragments which can still clot but which are more slowly coagulable. With streptokinase therapy one gets much smaller further degradation products of fibrinogen.

H.J. Röthig

Couldn't the high unphysiological doses of t-PA used in therapy induce the production of antibodies that would neutralize the effect of t-PA if it had to be used later, in the case of a second infarction?

H.R. Lijnen

The doses of t-PA used in therapy are very high, about 500-fold above the physiological level, but to my knowledge t-PA administration has never been associated with generation of antibodies. On the other hand, there is very little information of the repeated use of t-PA in patients.

© 1991 Elsevier Science Publishers B.V. (Biomedical Division)
The clinical pharmacology of biotechnology products.
M.M. Reidenberg, editor

ERYTHROPOIETIN.

FERNANDO VALDERRABANO.
Servicio de Nefrología. Hospital General Gregorio Marañón. Madrid. Spain.

The frequent occurrence of anemia in patients with renal failure is well known since the first classical paper on kidney disease from Richard Bright in 1836. However, the role of the kidney in regulating erythropoiesis was not recognised until 1957 when Jacobson and coworkers showed that removal of kidneys nearly abolishes erythropoietin (EPO) production in rats (1).

Subsequent studies showed that the normal isolated perfused kidney synthesizes EPO when perfused with a serum free medium and that this synthesis is augmented when the kidney is perfused at low pO2 (2).

The kidney is responsible for sensing oxigen availability to tissues, as well as for releasing EPO into circulation. The site of EPO generation within the kidney is a matter of debate. Peritubular interstitial cells, mesangial cells and yuxtaglomerular apparatus have all been considered as possible sites of renal EPO generation (3). Hepatic generation of EPO contributes only 10 to 15 % of total EPO in normal adults.

FUNCTION OF ERYTHROPOIETIN

EPO induces bone marrow red cell formation by stimulating proliferation, downstream differentiation and maturation of erythroid progenitors and precursors. In high titers EPO causes early release of reticulocytes into circulation (4). The most primitive erythroid progenitor cell responding to EPO is the burst forming unit-erythroid (BFU-E), and the more mature, the colony forming unit-erythroid (CFU-E). Proliferation of BFU-E requires a specific growth factor generated by lymphocytes or monocytes, and BFU-E will die if further differentiation does not occur.

BFU-E are minimally responsive to EPO, and CFU-E is more sensitive and specific for EPO action. EPO binds to receptors on the CFU-E, stimulating the proliferation and differentiation of these cells into erythroblats (5, 6).

Some studies show the existence of two classes of binding sites on erythroid progenitor cells, a high affinity and a low affinity EPO receptor. The high affinity receptor seems to be essential in the response to EPO. It is unknown how the receptors modulate the action of EPO. However, in the early stage of in vitro growth CFU-E have an almost absolute need for EPO (7, 8).

Although this early need for EPO by CFU-E is almost absolute, the often large amount of EPO already internalized into the cell cannot satisfy this need. This implies that CFU-E require repeated occupancy of the rapidly turning over EPO receptors. Thus, a more or less continuous presence of EPO is necessary to maintain the development, maturation and differentiation of red cells. In vivo, the biological effect of a given dose of EPO is largest when it is administered intermittently in divided gifts. These studies have been made using recombinant human erythropoietin (r-HuEPO) and they might explain some of the complex therapeutic properties r-HuEPO exhibits.

The sequence of biochemical postreceptors events tiggered by EPO is still undefined. More is known about the later events that occur as the erythroid progenitor cells mature into late erythroblasts. At the cellular level EPO enhances Ca^{++} uptake into the cells. Total RNA synthesis increases due to the activation of transcription within 3 to 4 hours and does not apprear to require protein or DNA-synthesis. A marked hemoglobin synthesis follows within 12 hours.

PATHOGENESIS OF THE ANEMIA OF CHRONIC RENAL FAILURE (CRF).

The most important cause of the anemia of CRF is decreased EPO production from the diseased kidney, but other factors contribute to the pathogenesis of the anemia.

Decreased red cell survival.

Decreased red cell survival to approximately half normal is seen in patients with advanced renal insufficiency (9). However, hemolysis is mild enough that a normal hematopoietic system should be able to compensate for it. Hemolysis is caused by substances in uremic plasma that interfere with the red blood cells membrane's ability to effectively pump sodium from the cells.

Inhibition of erythropoiesis.

Several lines of evidence suggest that uremic toxins retained in patients with CRF interfere with bone marrow function (6, 9). Uremic sera inhibit proliferation of BFU-E and CFU-E, hemo synthesis and bone marrow thymidine incorporation. Several compounds have been suggested to be the specific erythroid inhibitor including spermine, parathyroid hormone (PTH) and ribonuclease (9), but recent studies suggest that uremic inhibition of in vitro hematopoiesis is not specific.

Erythropoietin deficiency.

The anemia of CRF is primarily due to decreased EPO secretion by the diseased kidney since serum EPO levels are generally inappropriately low for the degree of anemia. With the purification of human EPO valid radioimmunoassays (RIA) for EPO were developed, and using these methods, mean values for serum EPO were 17.2 mU/ml for normal males, 18.8 mU/ml for normal females, 97 mU/ml for iron defficiency anemia patients, and greater than 1000 mU/ml for patients with anemia secondary to bone marrow failure (10). In our own experience, using the RIA we have found normal or high serum EPO levels in CRF patients, but they are low in relation to the degree of anemia of these patients. The usual inverse correlation between hematocrit or hemoglobin and serum EPO levels as would be expected if renal function were normal is not seen in patients with CRF.

Several studies have shown that in CRF patients serum EPO levels do not change with the decline in GFR despite the development of severe anemia (6). However, other authors recently showed a rise in serum EPO in hemodialysis patients in response to spontaneous hemorrhage, and suppression in serum EPO levels following blood transfusion (11).

The available evidence suggest that the hypoxia-EPO-hematocrit feedback system functions at a lower set point of tissue oxigenation in patients with renal failure than in normal individuals.

Other factors.

Blood loss may also contribute to the anemia in patients with advanced renal failure, as consequence of platelet dysfunction. Repetitive blood loss may cause iron deficiency.

Aluminium toxicity, hyperparathyroidism, folate deficiency, hypersplenism and acute hemolysis are aggravating factors of the anemia of CRF.

RECOMBINANT HUMAN ERYTHROPOIETIN.

Miyake and coworkers in 1977 isolated a large quantity of EPO from the urine of severely anemic patients and this material served as a source of pure EPO for the study of its chemical structure (12).

EPO is a glycoprotein with a molecular weight of approximately 34,000 daltons and contains approximately 25 % carbohydrate consisting mostly of sialic acid (6). Removal of the sialic acid portion of the molecule abolishes the in vivo but not the in vitro, activity of EPO. The carbohydrate moiety is not critical for the erythropoietic action of EPO, but prevents its rapid clearance.

The protein is made up of 166 aminoacids. Part of the aminoacid sequence was used to predict the base composition of a corresponding cDNA, wich in turn was used as a probe to identify the entire EPO gene. The EPO gene is encoded as a single copy on chromosome 7.

Molecular biologists have isolated the gene and inserted it into mammalian cells capable of synthesizing unlimited quantities of EPO. It was necessary to use animal cells rather than bacteria because EPO is highly glycosylated, and only animal cells provide the necessary sugar components.

The biological activity and immunological properties of human purified EPO and recombinant human EPO have not been distinguishable.

The first published reports of the successful isolation, cloning and expression of the human EPO gene appeared in 1985 (13, 14), and large amounts of r-HuEPO became available for clinical trials. The results of these trials involving anemic patients on hemodialysis from the basis of multicenter trials in the United States, Western Europe and Japan, coupled with the results in anemic patients with CRF not yet on dialysis, support the concept that EPO deficiency is the major mechanism responsible for the anemia (15, 16).

These studies showed that intravenous EPO therapy could fully reverse the anemia of CRF, and that there was a dose-dependent rate of response to EPO.

THERAPEUTIC USE OF r-HuEPO.

Large multicenter trials have extended and confirmed the initial observations with r-HuEPO. Virtually all patients treated responded appropiately to a intravenous dose of 50 to 300 U/kg, three times a week and increased their hematocrit to the target range. The patients no longer needed transfusions and their quality of life improved. Lower doses and differents routes of administration have been explored with similar good results.

Which patients should be treated?

Patients who are symptomatic of anemia should be treated. But it may be difficult to define who is symptomatic. Patients requiring blood transfusions should be treated. In the non-transfusion dependent patient, the following symptoms are associated with anemia and improve with partial correction of the anemia: physical fatigue, poor appetite, coldness, disordered sleep/awake pattern, depression, sexual desinterest and mental slowness. These symptoms are often present with a hemoglobin < 8 g/dl. However, dialysis patients with symptoms of ischemic heart disease, even if their hemoglobin is above 8 g/dl, sould also be considered for EPO therapy.

Patients who have recently commenced dialysis may experience spontaneous amelioration of their anemia to more acceptable levels, thus negating the need for EPO.

For patients with CRF not yet requiring dialysis the situation is even less clear. Few of these patients have a hemoglobin level < 8 g/dl and if they are symptomatic, EPO therapy may be appropiate. The increased blood viscosity resulting from the improved hematocrit may adversely affect renal perfusion and this accelerate the decline in renal function, although the clinical studies to date do not support this hypothesis (17, 18).

In all patients, any other treatable cause of anemia must be excluded before starting EPO therapy.

Dosage of EPO and routes of administration.

A variety of dosage regimes and routes of administration have been employed. The greatest experience is with intravenous therapy in hemodialysis patients and the earliest studies showed a dose-dependent rate of response to EPO. However, the risk of side effects such as severe hypertension and thrombotic complications is lessened with a hemoglobin rise not exceeding 1 g/dl/month.

As consequence the recommended starting dose of EPO has declined in comparison with earlier studies. Most centers now use an initial intravenous dose in the range of 40-50 U/kg three times a week, for hemodialysis patients. A similar IV dosage regime has been used with good results in patients not yet on dialysis (17).

The IV route is impractical for regular use in CAPD patients. Obvious alternatives to be considered include the intraperitoneal and subcutaneous routes. An effective clinical response has been obtained in these patients with a dose of 300 U/kg/week using the intraperitoneal route, but a similar response has been obtained with only 120/U/kg/week when EPO is given subcutaneously (17).

In the other hand, it has been shown that a 50 % reduction in dose can be achieved at optimal hemoglobin level by switching from IV to SC administration, in hemodialysis patients (19).

Thus, the SC route appears to be gaining popularity, not only in CAPD patients but also in hemodialysis and pre-dialysis patients, and evidence to date suggest that lower doses of EPO may be used when given by this route (60-150 U/kg/week). If the patient can be taught to give their own SC injection without stress or discomfort, then the daily dosing regime may be worth considering (14 U/kg/day) (20).

Pharmacokinetics.

Following IV administration of EPO, plasma concentrations reach a high peak shortly after the injection, with a mean T 1/2 ranging from 4.9 to 9.3 hours (7, 21). In most studies, the apparent volume of distribution equaled the plasma volume. The liver is considered the most likely site of degradation of EPO, but the bone marrow may make a contribution to EPO degradation.

After SC administration plasma concentrations of EPO start to increase after 2 hours. Peak concentrations are found at 12 to 18 hours and are much lower when compared to IV administration, but they remain above baseline for up to 72 hours (7, 21).

Target hemoglobin and rate of rise.

It is possible to fully correct the anemia of CRF with EPO. However, in comparing the benefits with the risks, partial correction of the anemia seems the best compromise. A linear increase in the hematocrit leads to an exponential rise in whole blood viscosity, which in turn is thought to contribute to many of the side effects of EPO therapy, such as hypertension and thrombotic complications.

The optimum hemoglobin seems to be in the range of 10-12 g/dl. Nevertheless, this is a very arbitrary guideline, and some flexibility is necessary in treating individual patients. Since the main aim of EPO therapy is to reverse the symptoms of anemia, differing thresholds at which this occurs may influence the appropriate target hemoglobin.

With regard to the rate of rise of the hemoglobin response, an increase of 1 g/dl/month should not be exceeded.

THERAPEUTIC EFFECTS OF EPO.

Correction of anemia in CRF patients results in better tissue oxigenation. The clinical benefits of this improved oxigenation include improved exercise tolerance, skin circulation and central nervous system function.

Tha almost constant feeling of tiredness is relieved, so that most patients are able to increase their activities, either in their work or their social life. There are also reports of increased appetite, relief of Raynauds' phenomenon, improvement of angina and of increased libido.

Exercise tolerance has been examined by several groups, showing a significant improvement. Working capacity, maximal oxigen consumption and anaerobic threshold increase after EPO therapy. A significant progresive reduction in left ventricular mass, measured using echocardiography, and a consequent improvement in cardiac function has been noted by different authors (17).

A number of investigators using questionnaires have now provided evidence of improvement of quality of life. An improvement in cognitive function with EPO therapy has been suggested. Improvement in sexual performance after 4 months of EPO therapy has been related with normalization of prolactin levels in these patients (22).

The effect of EPO on iron metabolism.

With the response to EPO, erythropoiesis may be stimulated as much as 3 to 4 times normal. This imposes a demand to movilize iron from reticuloendothelial storage sites and to make it available to transferrin for transport to the marrow for incorporation into hemoglobin. Severely anemic individuals who begin therapy with EPO are at risk of becoming iron deficient during the course of therapy. Currently, recommendations for managing patients during the induction phase of therapy include oral iron supplements, and in some patients IV administration of iron dextran.

In some patients the response to EPO is so brisk that movilization of iron from storage sites cannot keep place with the demand. In this setting, EPO may become less efective despite the fact that serum ferritin values clearly indicate that adequate iron stores are present. This condition represents a "functional" iron deficiency. The dose of EPO may be reduced or supplemental iron may be given orally or intravenously.

ADVERSE EFFECTS

The major adverse effect of raising hematocrit is aggravation of pre-existing hypertension, related to changes in peripheral vascular resistance. Peripheral vascular resistance increases as consequence of the increase of whole blood viscosity and by relief of hypoxia-induced peripheral vasodilation.

Although there is usually a normalisation of the elevated cardiac output as the anemia is corrected, occasionally this does nor occur, and a sustained high cardiac output would result in an increase in blood pressure (17).

A further major complication of EPO therapy is thrombosis of the arteriovenous fistula. This tendency for EPO treatment to increase the risk of thrombosis suggested that it may have a beneficial effect on the bleeding diathesis associated with uremia.

Less frequent side effects are flu-like symptoms, fever, bone pain, myalgia and seizures.

Table I shows the frequency of different adverse EPO effects from an European Multicenter Study (23), in 150 patients with a follow-up of one year. Adverse effects were arbitrarily classified as related to the hematocrit increse, to the drug itself and as concomitant events.

T A B L E   I
EPO ADVERSE DRUG EFFECTS

1.- Related to the hematocrit increase:

Hypertension	32 %
Clotting	12 %
Thrombosis of the fistula	14.7 %
Weight gain	10 %
Hyperkaliemia	7.3 %

2.- Related to the drug:

Musculoskeletal pain	30 %
Headache	26.6 %
Pyrexia	13.3 %
Flu-like syndrome	6.6 %
Pruritus	12 %
Rash	2.6 %

3.- Concomitant events:

Seizures	1.9 %
Cerebrovascular accidents	1.3 %
Myocardial infarction	1.3 %
Sudden death	0.6 %

Most of the adverse effects seems to be related to the hematocrit increase. In consequence, the aim of the treatment should be to increase the hamatocrit gradually by the use of an appropriate dosage regime of EPO with a view to reducing their incidence.

Genuine intolerance to EPO sufficient to warrant stopping the hormone is absolutely rare. There have been no reports of antibody formation.

EPO RESISTANCE

The large multicenter trials of EPO indicate that 95-98 % of patients treated with EPO will respond.

Potential causes of EPO resistance are iron deficiency, B12/folate deficiency, aluminium toxicity, hyperparathyroidism, infection, malignancy, blood loss and hemolysis.

Thus, patients with a poor response to EPO , or loss of a previous response, require investigation for an underlying cause.

NEW INDICATIONS

The efficacy of EPO is now being evaluated in non renal causes of anemia, such as rheumatoid arthritis, malignancy, thalassemia, sickle cell disease, myelofibrosis, etc. Preliminary reports show the efficacy of EPO therapy in anemic AIDS patients treated with AZT.

CONCLUSION

EPO replacement therapy wiyh r-HuEPO has been one of the most dramatic medical advances and the most clinically important achievement of the last decade in the treatment of chronic renal failure patients. EPO therapy demonstrates that the primary cause of anemia of uremia is a deficiency in renal EPO and that many symptoms of uremic syndrome are consequence of the associated anemia. Substancial reversal of anemic symptoms and quality of life improvement is possible when anemia is corrected with EPO therapy.

REFERENCES
1.- Jacobson LO, Goldwasser E, Fried W, Pizak L: The role of the kidney in erythropoiesis. Nature 179: 633. 1957.

2.- Erslev AJ: In vitro production of erythropoietin by kidneys perfused with serum free solution. Blood 44: 77. 1974.

3.- LaCombe C, DaSilva JL, Brunevd P, Fowinier JG, Wendling F, Casadivall N, Camilleri JP, Bariety J, Varet B, Tambowin P: Peritubular cells are the site of erythropoietin synthesis in the murine hypoxic kidney. J Lab Invest 81: 620. 1988.

4.- Spivak JL: The mechanism of action of erythropoietin. Int J Cell Cloming 4: 139. 1986.

5.- Sawyer ST, Krantz SB, Goldwasser E: Binding and receptor mediated endocytosis of erythropoietin in Friend virus infected erythroid cells. J Biol Chem 262: 5554. 1987.

6.- Chandra M: Pathogenesis of the anemia of chronic renal failure: The role of erythropoietin. Nefrología 10, supl 2: 12. 1990.

7.- Frenken LAM, Koene RAP: Recombinant human erythropoietin and the effects of different routes of administration. Nefrología 10, supl 2: 33. 1990.

8.- Landschulz KT, Noyes AN, Rogers O, Boyer SH: Erythropoietin receptors on murine erythroid colony-forming units: natural history. Blood 73: 1476. 1989.

9.- Eschbach JW, Adamson JW: Anemia in renal disease. In Diseases of the kidney. Editors: Schrier RW, Gottschalk CW. Little, Brown and Co. Boston, Toronto 1988. p.3019.

10.- García JF, Ebbe SN, Hillander L, Cutting HO, Miller M, Cronkite EP: Radioimmunoassay of erythropoietin. Circulating levels in normal and polycythemic human beings. J Lab Clin Med 99: 624. 1982.

11.- Walle AJ, Wong Y, Clemons GK, García JF, Niedermayer W: Erythropoietin hematocrit feed-back circuit in the anemia of end stage renal disease. Kidney Int 31: 1205. 1987.

12.- Miyake T, Kung CK-H, Goldwasser E: Purification of human erythropoietin. J Biol Chem 252: 5558. 1977.

13.- Jacobs K, Shoemaker C, Rudersdorf R, Neill SD, Kaufman RJ, Musfson A, Seehra J, Jones SS, Helwick R, Fritsch EF, Kawakita M, Shimizu T, Miyake T: Isolation and characterization of genomic and cDNA clones of human erythropoietin Nature 313: 806. 1985.

14.- Lin FK, Suggs S, Lin CH, Browne JK, Smalling R, Egrie JC, Chen KK, Fox GM, Martin F, Stabinsky Z, Badrawi SM, Lai PH, Goldwasser E: Cloning and expression of the human erythropoietin gene. Proceedings of the National Academy of Sciences USA 83: 6465. 1986.

15.- Winearls CG, Oliver DO, Pippard MJ, Reid C, Downing MR, Cotes PM: Effect of human erythropoietin derived from recombinant DNA on the anemia of patients maintained by chronic hemodialysis. Lancet 2: 1175. 1986.

16.- Bommer J, Kugel M, Schoeppe W, Brunkhorst R, Samtleben W, Bramsiepe P, Scigalla P: Dose-related effects of recombinant human erythropoietin on erythropoiesis: results of a multicenter trial in patients with end stage renal disease. Treatment of renal anemia with Recombinant Human Erythropoietin. Contr Nephrol 66: 85. 1988.

17.- Macdougall IC, Hutton RD, Cavill I, Coles GA, Williams JD: Recombinant human erythropoietin in the treatment of renal anemia: An update. Nefrología 10, supl 2: 23. 1990.

18.- Lim VS, DeGowin RL, Zavaba D, Kirchner PT, Abels R, Perry P, Frangman J: Recombinant human erythropoietin treatment in predialysis patients: a double-blind placebo-controlled trial. Ann Intern Med 110: 108. 1989.

19.- Bommer J, Ritz E, Weinreich T, Bommer G, Ziegler T: Subcutaneous erythropoietin. Lancet 2: 406. 1988.

20.- Granolleras G, Branger B, Beau MC, Deschodt G, Alsabadani B, Shaldon S: Experience with daily self-administered subcutaneous erythropoietin. Contr Nephrol 76: 143. 1989.

21.- Stevens JM, Winearls CG: Clinical use of recombinant human erythropoietin in hemodialysis and CAPD patients. Nefrología 10, supl 2: 38. 1990.

226

22.- Schaefer RM, Kokot F, Kürner B, Zech M, Heidland A: normalization of elevated prolactin levels in hemodialysis patients on erythropoietin. Nephron 50: 400. 1988.

23.- Valderrábano F: Adverse effects of recombinant human erythropoietin in the treatment of anemia in chronic renal failure. Nephrol Dial Transpl 3: 503. 1988.

Due to unavoidable circumstances, the discussion after this presentation could not take place.

© 1991 Elsevier Science Publishers B.V. (Biomedical Division)
The clinical pharmacology of biotechnology products.
M.M. Reidenberg, editor

CLINICAL PHARMACOLOGY OF HIRUDIN (HBW 023)

H.-J. ROETHIG, J.S. MAREE, B.H. MEYER
Hoechst AG, Department of Clinical Pharmacology, P.O. Box 800320, D-6230
Frankfurt (M) 80 and University of the Orange Free State, Department of
Pharmacology, P.O. Box 339, Bloemfontein 9300, Republic of South Africa

INTRODUCTION
Leeches are small fresh water animals sucking blood from other animals or
man. The capacity to suck blood is enormous. After an opulent meal,
leeches can remain without food for many months. This would not be possible without a mechanism that prevents blood from clotting during and after
the sucking process. In 1884, John Haycraft discovered that leeches secrete a substance which inhibits blood clotting (1). This substance was
later on called hirudin. It is a single chain polypeptide with a molecular
weight of 7000 daltons, and it consists of 65 amino acids (Figure 1) (2).
Hirudin binds specifically to thrombin with a very high affinity (3). This
is surprising because thrombin is only a slightly modified trypsin like
proteinase. The carboxy terminal end of the hirudin molecule binds near
the active centre of thrombin and other parts of the hirudin molecule to
other sites of the thrombin molecule . So far, no other enzymes were found
to be inhibited by hirudin. There are a few reports of additional actions
of hirudin: acceleration of the displacement of Factor Xa from the vascular
endothelium and association with Factor Xa (4). Hirudin is a rather
stabile peptide, quite resistant against proteinase and heat degradation
and well soluble in water.

MATERIALS AND METHODS
Hirudin (HBW 023) was produced by r-DNA technology in yeast cells. After
purification, the product had a purity of more than 97 %. Special investigations were undertaken to assure the quality of the product (Table 1).
The product was supplied in vials containing 10 mg of lyophilyzed hirudin
to be dissolved in 1 ml of physiological saline.

In rats, dogs and monkeys, hirudin showed dose-dependent increases of
thrombin time and activated partial thromboplastin time after i.v. and s.c.
application in doses of 0.01 - 10 mg/kg. In single dose toxicological
studies the LD_{50} could not be determined due to good tolerance. In repeated dose administration over 4 weeks and 3 months in rats and monkeys

Table 1

Quality Control of Hirudin (HBW 023)

Peptide Characterisation	
Biologic Activity	Thrombin Inhibition
Peptide Sequence	Amino Acid Analysis
Molecular Weight	SDS-Gel Electrophoresis
Molecular Charge	Isoelectric Focussing
Molecular Size	Gel Permeation Chromatography

Table 2

Phase I Program of Hirudin (HBW 023)

Single Dose Studies	0.01	0.025	0.05	0.07 mg/kg i.v.
	0.1	0.2	0.3	0.5 mg/kg i.v.
Single Dose Studies	0.05	0.1	0.15	mg/kg s.c.
	0.2	0.35	0.5	mg/kg s.c.
Multiple Dose Studies	5 x 0.1 mg/kg		every 24 hours i.v.	
	5 x 0.1 mg/kg		every 12 hours i.v.	
	5 x 0.5 mg/kg		every 24 hours s.c.	
	5 x 0.5 mg/kg		every 12 hours s.c.	

Table 3

Pharmacokinetics of Hirudin (HBW 023)

Means and standard deviations of 5 volunteers per dose group

Dose mg/kg	0.1	0.2	0.3	0.5	0.1	0.2	0.35	0.5
Route of administration	i.v.	i.v.	i.v.	i.v.	s.c.	s.c.	s.c.	s.c.
C_{max} (ng/ml)	859* 106	1177* 133	1818* 219	3443* 327	125 12.5	231 47.2	346 108	382 61.3
AUC (ng·h/ml)	652 118	1098 111	1562 351	2959 473	498 65.2	1098 188	1695 234	2064 249
t_2 (h)	0.89 0.15	1.16 0.07	1.10 0.29	1.15 0.18	1.80 0.77	1.92 0.14	2.31 0.78	2.78 0.92
Vss (l)	14.3 3.08	22.4 3.57	19.0 2.01	20.5 3.37				
Cl-tot (ml/min)	205 44.1	241 30.2	254 56.3	227 28.3	257 28.3	241 36.7	236 38.6	269 31.8
Cl-ren (ml/min)	94.2 17.8	90.4 15.2	110 30.9	115 14.7	97.6 15.4	73.5 10.2	- -	- -
Ae (% of dose)	49.4 8.10	38.6 5.64	44.0 5.60	51.8 4.80	43.7 5.77	35.4 1.86	- -	- -

* First observed concentration

Figure 1

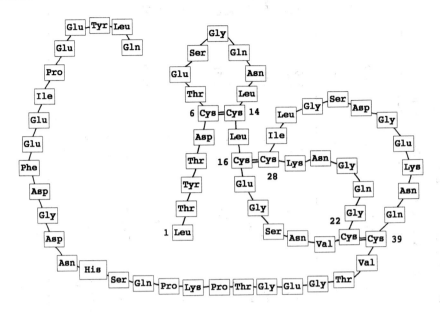

Figure 2

**Single dose kinetics of Hirudin (HBW 023) intravenously
Hirudin plasma concentrations
Mean values (n=5)**

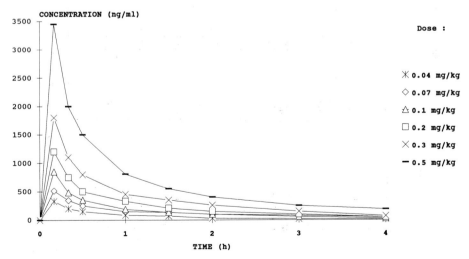

intravenously and subcutaneously, no toxic effects could be observed, except the anticoagulant effect of hirudin with higher doses. The local tolerance of hirudin was also good. There were also no hints for mutagenic effects.

According to Blackwell (5), the first dose in man was calculated to be 0.01 mg/kg.

For the determination of hirudin in plasma and urine, we used a thrombin inhibition assay (6). In order to have adequate precision and sensitivity, individual calibration curves had to be established for each volunteer from his plasma blanks. The detection limit of this assay is 9 ng/ml for plasma and 60 ng/ml for urine.

RESULTS

The Phase I program included single dose studies i.v. and s.c. and multiple dose studies i.v. and s.c. (Table 2). The results of single dose kinetic studies are given in Figure 2 for intravenously applied hirudin and Figure 3 for subcutaneously applied hirudin. The pharmacokinetic parameters are summarized in Table 3. Hirudin has dose-linear kinetics with a half-life of approximately 1 hour. Major route of excretion is renal elimination, probably mainly by glomerular filtration. It becomes evident that renal clearance accounts only for 50 % of the total body clearance. Around 30 to 50 % of the administered hirudin was found unchanged in urine. This is in contrast to former reports of natural hirudin and kinetics in animals where renal elimination was found to contribute more than 90 % to elimination (7).

After subcutanous administration, hirudin is well absorbed from the tissue with C_{max} values occurring around 2 hours. Comparison of AUC and urinary excretion with i.v. dosing indicates 80-90 % bioavailability.

In multiple dose studies intravenous doses of 0.1 mg/kg every 12 or every 24 hours for 5 times were well tolerated. The concentration time curves were rather identical for all days. So, no accumulation of hirudin occurred. The same holds true for subcutanous administration.

In total, more than 50 volunteers participated in the Phase I trials. Hirudin was very well tolerated, no side effects were noted. No specific antibody induction could be detected against hirudin or yeast proteins.

Figure 3

Single dose kinetics of Hirudin (HBW 023) subcutanously
Hirudin plasma concentrations
Mean values (n=5)

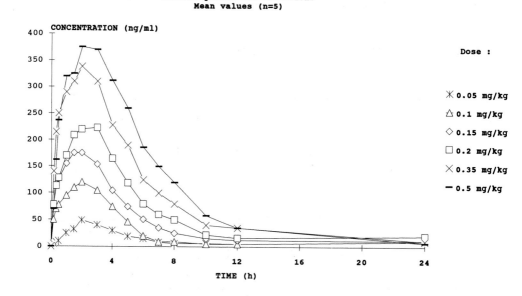

Dose :

✳ **0.05 mg/kg**

△ **0.1 mg/kg**

◇ **0.15 mg/kg**

☐ **0.2 mg/kg**

✕ **0.35 mg/kg**

— **0.5 mg/kg**

Figure 4

Single doses of Hirudin (HBW 023) intravenously
Activated partial thromboplastin time (aPTT)
Mean values (n=5)

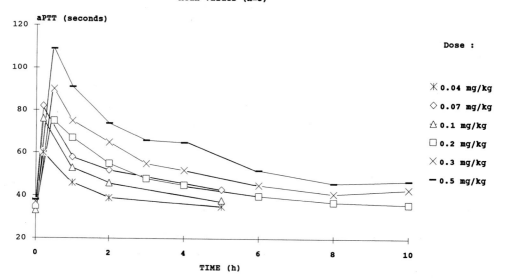

Dose :

✳ **0.04 mg/kg**

◇ **0.07 mg/kg**

△ **0.1 mg/kg**

☐ **0.2 mg/kg**

✕ **0.3 mg/kg**

— **0.5 mg/kg**

Figure 5

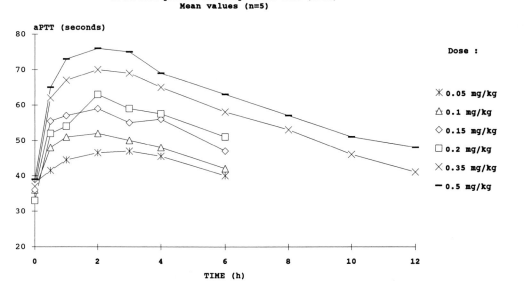

**Single doses of Hirudin (HBW 023) subcutanously
Activated patial thromboplastin time (aPTT)
Mean values (n=5)**

Dose :

✕ **0.05 mg/kg**
△ **0.1 mg/kg**
◇ **0.15 mg/kg**
☐ **0.2 mg/kg**
✕ **0.35 mg/kg**
— **0.5 mg/kg**

Figure 6

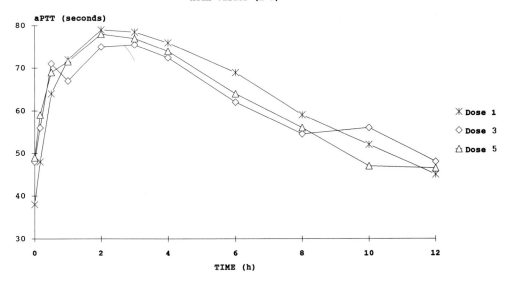

**Multiple doses of Hirudin (HBW 023)
Activated partial thromboplastin time (aPTT)
Mean values (n=5)**

✕ **Dose 1**
◇ **Dose 3**
△ **Dose 5**

234

Figure 7

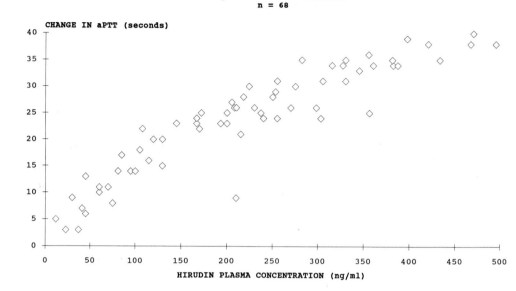

Pharmacodynamics were measured by several assays. As pharmacodynamic variables for in vivo effects we used bleeding time, thrombin time and activated partial thromboplastin time (aPTT). There were no effects, even not
in the highest dose on bleeding time. The thrombin time showed a very steep
dose response, so that for most of the time points the values were above
100 seconds. Thus, the thrombin time does not seem to be an appropriate
method for the pharmacodynamic measurement of hirudin. The effect of single
doses of hirudin on aPTT can be seen in Figures 4 and 5. After multiple
dosing i.v., the response on aPTT was identical for doses 1, 3 and 5
(Figure 6). Similar results were seen after multiple doses s.c.

The effect of hirudin was clearly related to plasma levels. After a linear
increase with low levels, the curve flattens with higher levels (Figure 7).
This is in contrast to the effects seen with other anticoagulants (8). The
effect of hirudin on aPTT can be described by the formula: aPTT = A-m/log
concentration of hirudin.

Hirudin application had no influence on the spontaneous thrombin generation. This can be concluded from constant F_{1+2} levels as marker of thrombin generation. But hirudin competes with AT III for thrombin as can be
seen by slightly decreasing levels of AT III/thrombin complexes during
hirudin therapy.

DISCUSSION

Haemostasis in blood is normally well regulated in man through a number of systems including activators, inhibitors and feed-back magnification. Nevertheless, under certain circumstances situations occur when the clotting of blood leads to disease rather than being beneficial. In these cases, anticoagulants are needed. But they all bear the risk of bleeding depending on the therapeutic range of the drug. Heparin is a widely used drug with well-proven clinical efficacy. Unfortunately, heparin is a mixture of molecules with different affinities for binding; heparin has multiple sites for binding, some are specific, others are unspecific. Heparin also binds to many proteins in blood and can also induce thrombocyte aggregation. The kinetics of heparin are non-linear and influenced by liver and kidney function. The individual anticoagulative response is very variable. The dynamic action can heavily be disturbed by histidin-rich protein. Patients with thrombosis need more heparin than healthy volunteers. The non-linear kinetics are accompanied by a poor correlation of plasma levels and pharmacodynamic effect. Thus it can be concluded that heparin is an anticoagulant with a very narrow therapeutic range (9).

In contrast, for a very specific thrombin inhibitor like hirudin a much wider therapeutic range is anticipated. From the data presented here we can see that the kinetics are linear with low variability and are very predictable. In contrast to heparin, hirudin effects correlate well with plasma levels. The effect is clearly dose-related. The flattening of the dose-response curve with higher hirudin levels might be a major point guaranteeing safety because overdosing would not necessarily mean bleeding. This is in contrast to heparin where the dose-response curve is much steeper with higher doses.

In conclusion, the presented Phase I data support the hypothesis that hirudin by its high specificity is a valuable anticoagulant which seems to have a much wider therapeutic range as compared to heparin and other anticoagulants. The safety profile of hirudin therapy is expected to be better than that of heparin.

REFERENCES

1. Haycraft JB (1884) On the action of secretion obtained from the medicinal leech on the coagulation of the blood. Proc Roy Soc London 36:478-87

2. Markwardt F, Walsmann P (1967) Reindarstellung und Analyse des Thrombininhibitors Hirudin. Z Physiol Chem 348:1381-6

3. Markwardt F (1957) Die Isolierung und chemische Charakterisierung des Hirudins. Z Physiol Chem 308:147-56

4. Friedberg RC, Hagen PO, Pizzo SV (1988) The role of endothelium in factor Xa regulation: The effect of plasma proteinase inhibitors and hirudin. Blood 71:1321-8

5. Blackwell B (1972) For the first time in man. Clin Pharmacol Ther 13:812-23

6. Griessbach U, Stuerzebecher J, Markwardt F (1985) Assay of hirudin in plasma using a chromogenic thrombin substrate. Thromb Res 37:347-50

7. Markwardt F, Nowak G, Stuerzebecher J, Griessbach U, Walsmann P, Vogel G (1984) Pharmacokinetics and anticoagulant effect of hirudin in man. Thromb Haemostas 52:160-3

8. Barrowcliffe TW (1989) Heparin Assays and Standardization. In: Lane DA, Lindahl U (eds) Heparin, Chemical and biological properties, clinical applications. Edward Arnold, London, pp 393-416

9. Albada H, Nieuwenhuis HK, Sixma JJ (1989) Pharmacokinetics of Standard and Low Molecular Weight Heparin. In: Lane DA, Lindahl U (eds) Heparin, Chemical and biological properties, clinical applications. Edward Arnold, London, pp 417-32

Discussion - CLINICAL PHARMACOLOGY OF HIRUDIN (HBW 023)

J.A. Galloway

Does hirudin have any antimicrobial effect? Because one rarely sees infections in people who have leeches attached to them.

H.J. Röthig

So far, I don't think so. Hirudin is a highly specific thrombin inhibitor. It has been assumed that it may also act on factor Xa, but this is unclear.

M.M. Reidenberg

You mentioned that half life of subcutaneously administered hirudin is more prolonged than that of IV hirudin. I tend to think of half life as being related to elimination and I am curious if you think the subcutaneous route actually modified the elimination of the drug or if the prolonged half life was simply an artifact caused by continuing absorption.

H.J. Röthig

Sure, it is an effect of retarded absorption. It has nothing to do with elimination, but it is clinically relevant. The interesting thing is that the absorption seems to be slower with increasing doses. At present, we haven't found out if this is a result of a reduced perfusion of the tissue, but that could be an explanation.

P. Tanswell

In your healthy volunteer studies you used the activated partial thromboplastin time as an end point. I would be interested to know what the end point was in your phase II studies, because I would imagine you would be looking for an antithrombotic effect rather than an effect on the coagulation system, and there you may not see such a clear cut relationship between plasma levels and pharmacological effect.

H.J. Röthig

We look for the prevention of thrombosis, that is the end point. There are certain clinical models, such as hip replacement surgery. There is a high percentage of thrombosis and that can be used in the quantification of therapeutic effects.

R.G. Werner

You mentioned that there are natural variants of hirudin and that there are also second generation hirudins. Is there any improvement in specificity or activity compared to the natural compound?

H.J. Röthig

Actually not. This molecule has such a high affinity to thrombin that by changing the molecule one can no longer improve that. Any reported difference on binding constants lies in the methodology used.

D.C. Brater

I want to make a comment concerning pharmacokinetics. The data that you already have could be used to answer the question as to whether or not the prolonged half life following subcutaneous administration is due to absorption. You can look at the mean residence time after the subcutaneous administration and subtract the mean residence time after intravenous administration and that gives you a parameter called absorption time. If that absorption time is greater than the mean residence time intravenously, then you have what is called "flip-flop kinetics" and the half life that you are actually looking at is the absorption half life, which is longer than the elimination half life. I also have a question: With heparin, if people get into trouble we have an antidote, that is protamine, but if someone who is anticoagulated with hirudin starts to bleed, what could you do to stop that?

H.J. Röthig

THARA is an antidote available. It is possible to administer activated factors like "autoplex" which is available in most countries. One can normalize the APTT and probably the bleeding tendency. In general, we think that the risk of bleeding with hirudin is not higher than with "aspirin".

M. Levy

Let me ask you a question concerning the antigenicity of hirudin. Did anyone look at people who were treated with leeches to see whether they formed antibodies?

H.J. Röthig

Yes, this was done and there one can find antibodies. When we first started we had this information and we spent a lot of time measuring possible formation of antibodies, particularly in our multiple dose subcutaneous studies. None of the volunteers developed specific antibodies against hirudin. We saw some unspecific effects but in our assay we can clearly discriminate between specific and non-specific antibodies.

© 1991 Elsevier Science Publishers B.V. (Biomedical Division)
The clinical pharmacology of biotechnology products.
M.M. Reidenberg, editor

THERAPEUTIC ACTIONS OF RECOMBINANT HUMAN GRANULOCYTE-MACROPHAGE COLONY-STIMULATING FACTOR (GM-CSF)

ARNOLD GANSER AND DIETER HOELZER
Department of Hematology, Johann Wolfgang Goethe-University,
Theodor-Stern-Kai 7, D-6000 Frankfurt 70 (Germany)

BIOLOGY OF GM-CSF

The colony-stimulating factors (CSFs) are glycoproteins which promote and modulate the proliferation and functional activity of various hematopoietic cell populations in vitro and in vivo (1,2). One member of this family is granulocyte-macrophage colony-stimulating factor (GM-CSF) which is a multilineage stimulator preferentially promoting growth and development of neutrophil, monocyte/macrophage and eosinophil progenitor and precursor cells (3-9). Depending on the concentration of GM-CSF in culture, the proportion of cells in cycle, their mean cycle times and the total number of progeny produced are increased (2). With increasing concentration, GM-CSF also becomes an effective stimulator of megakaryocytic and then some erythroid and multipotential progenitor cells (4,6). The effects of GM-CSF on the early progenitors can be enhanced by interleukin-3, interleukin-1, interleukin-6 and G-CSF (10-15).

Apart from its action on progenitor and precursor cells, GM-CSF activates mature, post-mitotic blood cells, increasing phagocytosis, cytotoxicity and superoxide generation by monocytes, neutrophilic and eosinophilic granulocytes as well as the production of prostaglandin-E, gamma-interferon, tumor necrosis factor, plasminogen activator and other CSFs by macrophages (7,16-25). In addition, GM-CSF inhibits neutrophil migration (26-28).

Human GM-CSF is a glycoprotein with a molecular weight of 15-30 Kd depending on the degree of glycosylation. The polypeptide is a single chain of 127 amino acids and a MW of 14.7 Kd (3,29). The gene encoding GM-CSF is located on the long arm of chromosome 5 in close proximity to other grwoth factor genes including interleukin-3, interleukin-4, interleukin-5 and M-CSF (30,31).

GM-CSF exerts its action through binding to high and low affinity GM-CSF receptors which are present in a small number on the membrane of responding hematopoietic cells (32-34). Membrane receptors for GM-CSF are not only present of hematopoietic cells, but also on leukemic cells leading to leukemic cell proliferation

(35-44), as well as on a number of nonhematopoietic cells, such as small-cell lung cancer, ovarian carcinoma and colon-carcinoma cell lines, normal fibroblasts and endothelial cells stimulating in vitro growth and function (45-48). The relevance of these effects in vivo, especially when using GM-CSF after chemo-/radiotherapy for malignant diseases, has not yet been defined.

Normally, GM-CSF is not detectable in the urine or plasma. However, after induction by agents like interleukin-1, endotoxin or foreign antigen, GM-CSF production is highly increased within hours. Sites of production are endothelial cells, fibroblasts, stromal cells, macrophages and T- and B-lymphocytes (49-56).

Recombinant DNA technology has allowed to produce recombinant human GM-CSF in sufficient amounts for preclinical and clinical trials. Human GM-CSF has been expressed in yeast (57,58), Escherichia coli (29) and COS cells (6,59). Carbohydrate moieties are not necessary for either receptor binding or activation, since in vitro and in vivo studies have shown full biologic activity of non-glycosylated GM-CSF (E. coli synthesized) and the partly-glycosylated GM-CSF (produced in yeast) as compared to the fully glycosylated GM-CSF (produced in mammalian cells), but might have some importance in the retardation of clearance and degradation.

IN VIVO EFFECTS OF GM-CSF
Pharmacokinetics

After IV bolus injection of recombinant GM-CSF, the initial phase of clearance is between 5-10 minutes (60-62) with a T1/2β of 85-150 minutes, assuming two phases of elimination. Peak serum concentrations of GM-CSF reached after IV bolus injection of 0.3-1.0 µg/kg or IV infusion over 30 minutes of 10-60 µg/kg range between 14-54 ng/ml and 35-135 ng/ml, respectively (60-62). After IV bolus injections, however, a stimulatory GM-CSF serum concentration >1ng/ml is only maintained for at least twelve hours, if high dosages are administered (62).

In contrast, GM-CSF serum concentrations >3ng/ml can be achieved by either IV continuous infusion of GM-CSF at a dosage of 3µg/kg/day (62) or by SC bolus injection. To obtain a serum concentration >1ng/ml for a prolonged period by SC bolus injection, a GM-CSF dosage above 3 µg/kg/day has to be given (60,63). After SC administration of 10-15 µg/kg, serum concentrations of 5-20 ng/ml are obtained within 2-6 hours which remain >1ng/ml, i.e. in a stimulatory range, for 12-24 hours(60,63,64).

Hematological Effects

In patients with malignant disease but not receiving chemotherapy, a significant stimulation of neutrophils, eosinophils and monocytes is observed (60,61,64,65). After an immediate but transient fall in circulating neutrophils, eosinophils and monocytes within 15-30 minutes of IV or SC GM-CSF administration (61,62,64,66), a leukocytosis of up to 10-fold with increases in numbers of circulating neutrophils, eosinophils, monocytes and lymphocytes is observed which is maintained throughout the treatment period but returns to baseline levels within one to two days after the end of GM-CSF application (61,63,64). Granulocytes and monocytes/macrophages are activated after in vivo administration of GM-CSF leading to increased phagocytic activity, monocytic CD11 expression, and release of secondary granules and superoxide anion from neutrophils (61,67-70), while the migration of neutrophils is inhibited (28).

There appears to be a plateau in the increase in neutrophils in the dose range of 3-15 µg/kg/day (71). Continuous IV or SC infusion or SC bolus infection of GM-CSF is more effective than rapid IV infusion (61,71). Optimal doses of GM-CSF in patients not receiving chemotherapy appear to be 250-500 µg/m²/day ($\approx$6-12 µg/kg/day), but even doses of 1000 µg/m²/day are tolerable, and doses up to 64 µg/kg ($\approx$2500 µg/m²/day) have been given (72).

Treatment with GM-CSF results in a dose-dependent increase in bone marrow cellularity and myeloid:erythroid ratio as well as in an increase in eosinophils (64). The cell cylce rate of hematopoietic progenitor cells in the bone marrow is increased within a few days (73-75). Although there is no change in the incidence of progenitor cells in the bone marrow of treated patients, the increase in bone marrow cellularity following treatment with GM-CSF implies that the absolute number of progenitor cells has increased. Similarly, after an initial decrease hematopoietic progenitor cells are recruited into the circulation by administration of GM-CSF, making them accessible for autologous transplantation (75-77).

Toxicity

Adverse effects of GM-CSF include mild fatigue, weakness,

fever, bone pain, anorexia, edema, transient dyspnea after the first dose and transient thrombocytopenia. At high dosages which are clinically unnecessary ($\approx$1000 μg/m²) GM-CSF can cause thrombosis, capillary leakage syndrome, effusions, respiratory distress and hypotension. Part of these adverse effects might result from the induction of secondary cytokines, like tumor necrosis factor, interleukin-1 or interleukin-6, and to the induced expression of cell adhesion molecules on leukocytes and endothelial cells (61,64,65,78). Rare adverse effects were the perforation of a granulocytic granuloma within the bowel walls (79). The development of anti-GM-CSF antibodies has been described although these were not neutrolizing (80). Their significance is not yet understood.

Clinical Use of GM-CSF

Treatment of cancer with myelotoxic agents, i.e. chemotherapy or large-field irradiation is accompanied by suppression of normal bone marrow function resulting in anemia, leukopenia or thrombocytopenia. Reversal of profound anemia and thrombocytopenia has usually required transfusion of red blood cells and platelets, while granulocytopenia could generally not be reversed. Especially profound neutropenia with cell counts <500/μl increases the risk of local infection and septicemia. Therefore, it has been general clinical practice to postpone or reduce the amount of myelotoxic chemotherapy in subsequent treatment cycles to prevent the occurence of cytopenia and related morbidity and mortality. This approach, however, increases the risk of a reduced effectiveness of tumorcidal therapy.

There now is an increasing number of studies addressing the issue of reduced myelotoxicity of cancer-chemotherapy by the use of GM-CSF. In general, they have demonstrated the beneficial effect of GM-CSF with tolerable adverse effects. In each of these trials, a dose-dependent decrease of neutropenia was observed with a shortening of the periods of neutropenia and intervals between cycles of chemotherapy (72,81,82). Despite being a multipotential hemopoietic growth factor, the effect of GM-CSF was mainly restricted to the granulocytic lineages with stimulation of the neutrophilic and eosinophilic lineages. However, in a controlled randomized trial (83) as well as two non-randomized trials (62,72) the degree of thrombocytopenia and the requirement for platelet transfusion were reduced in patients receiving GM-CSF. Stimulation of thrombopoiesis by SC injections of GM-CSF might require a change in the schedule, i.e. dividing the daily dose to two doses every 12 hours appears to be superior to a single daily dosage (84).

While in the non-randomized trial the rate of septicemia was not reduced by GM-CSF (72), less infectious complications (7% versus 24%) were experienced by the patients receiving GM-CSF in the until now only randomized trial (83).

GM-CSF was also used in patients with breast cancer or melanoma (85) and in patients with lymphoid malignancies (86,87) undergoing high-dose combination chemotherapy and autologous bone marrow transplantation. When compared with matched historical controls, leukocyte recovery was fastened, with fewer infections and earlier discharge from hospital. A more rapid platelet recovery, however, was only seen in one trial (86). Additional transfusion of circulating hemopoietic stem cells harvested during prior administration of GM-CSF can even more enhance hematological recovery after autologous bone marrow transplantation (76,88).

In patients with bone marrow failure after allogeneic bone marrow transplantation or unrelated donor transplantation, GM-CSF was effective in improving hematopoiesis and survival in comparison to age-matched controls (89). Similarly, after accidental radiation exposure GM-CSF has been successfully used to improve neutrophil recovery (90).

In a series of trials, GM-CSF has been given to patients with profound neutropenia and increased risk of severe infections, including patients with chronic idiopathic neutropenia (91-93), inherited neutropenia of childhood (94,95), AIDS (96), agranulocytosis (97), aplastic anemia (98-100) or myelodysplastic syndromes (63,64,99,101-105). In patients with chronic idiopathic neutropenia, treatment with GM-CSF has reversed neutropenia during the period of its administration leading to recovery from infections or preventing post-operative wound infections (91). The results in patients with agranulocytosis are diverse. In patients with known underlying mechanism, withdrawal of the offending agent and administration of GM-CSF has fastened neutrophil recovery, while in other patients in whom the underlying cause was unknown and therefore could not be influenced, GM-CSF failed to reverse neutropenia (97).

A rapid increase in neutrophil and eosinophil counts can be achieved by GM-CSF treatment in leukopenic AIDS patients (96). GM-CSF can also reverse or prevent drug-induced neutropenia which can be pronounced during antiviral therapy with azidothymidine (106) and gancyclovir (107). The modulation of HIV activity by GM-CSF has to be considered which can lead to a rise of p24 levels during therapy with GM-CSF alone. These findings indicate that GM-CSF should be combined with antiviral therapy whenever possible in these patients.

In patients with aplastic anemia, GM-CSF treatment has been successful only in patients with still some residual hematopoiesis in the bone marrow. While these patients responded to administration of GM-CSF with an albeit moderate increase in neutrophil counts (98,99), no response was observed in patients with very severy aplastic anemia, i.e. neutrophil counts below 200/μl (100). These findings indicate that a minimum number of stem cells have to be left in the bone marrow to obtain a hematopoietic response to the administration of GM-CSF.

Several clinical trials with rhGM-CSF in patients with myelodysplastic syndromes have been published to date including a total of 55 patients (63,99,101-104). Of these patients who were treated with different schedules, GM-CSF dosages and routes of administration, 84% showed a dose dependent increase in neutrophil counts. An increase in reticulocyte counts was observed in 38% of the patients. Platelets increased above baseline values in 15% of the patients. In eleven patients (20%), a transient increase in the marrow and/or peripheral blood blast cells was noted. Nine of the patients, particularly those with >15% bone marrow blasts, progressed to acute leukemia.

In the only randomized trial, patients with myelodysplastic syndromes received either 3 μg/m²/day GM-CSF SC or were observed (105). With more than 25 patients in each arm and treatment periods in excess of six months, all patients receiving GM-CSF had a sustained increase in neutrophil counts coupled with a decrease in infection rate. No effect was seen on the platelet or reticulocyte counts. Progression to acute leukemia was comparable in both arms totalling to about 10% each.

The substantial rise in neutrophils in patients with only abnormal metaphases in the bone marrow can be regarded as evidence that GM-CSF in some cases acted as an agent which in vivo can induce maturation of malignant myeloid cells (101,102). Premature chromosome condensation analysis of maturing granulocytes also indicate that the neoplastic precursor cells rather rather than normal hemopoietic progenitor cells are stimulated to differentiate (108). This is further supported by results from analysis of X-linked restriction fragment length polymorphism in a female patient with refractory anemia heterozygous for the X-chromosome linked gene pyruvate glycerol kinase (Ganser et al, submitted). Despite a response of the neutrophil counts to GM-CSF, the bone marrow and peripheral blood cells in this patients remained clonal. In contrast, Vadhan-Raj et al (109) recently published data showing that an individual patient with therapy-related myelodysplastic syndrome and pancytopenia achieved complete hematologic, cytogenetic and molecular genetic remission for nearly 1 year after discontinuation of GM-CSF.

As clinical studies in patients with myelodysplastic syndromes but increased blast cell load have shown, GM-CSF is capable of recruiting leukemic blast cells into proliferation in vivo. GM-CSF has therefore been used in combination with low-dose cytosin-arabinoside (110,111): first, because there is ample experience that low-dose cytosin-arabinoside alone can achieve responses in myelodysplastic syndromes (112), and second, because there is evidence for a synergistic effect with GM-CSF (113-117). Similarly, hematopoietic growth factors might be particularly useful for recruiting quiescent leukemic stem cells into cell cycle rendering them more sensitive to chemotherapeutic agents and increasing the log kill of the malignant clone. Preliminary results of ongoing trials demonstrate the feasability of this approach, but randomized trials will have to show whether the rate of complete remissions and the remission duration can be improved.

Conclusion

The initial data of the clinical trials suggest that GM-CSF is a potent stimulator of blood formation which can be used to alleviate chemotherapy/radiotherapy induced bone marrow suppression. Ongoing and future clinical trials will have to show whether the tumoricidal therapy can be dose-intensified to increase the response rates and remission duration periods. Future trials are also likely to use combinations of the growth factors to obtain multilineage hematopoietic responses.

REFERENCES

1. Groopman JE, Molina JM, Scadden DT: N Engl J Med 1989; 321:1449

2. Metcalf D: The Molecular Control of Blood Cells. Harvard University Press, London, 1988

3. Gough NM, Nicola NA. In: Dexter TM, Garland JM, Testa NG (eds) Colony-Stimulating Factors. Marcel Dekker Inc., New York, 1990; pp111-149

4. Sieff CA, Emerson SG, Donahue RE, et al: Science 1985; 230:1171

5. Emerson SG, Yang YC, Clark SC, Long MW: J Clin Invest 1988; 82:1282

6. Kaushansky K, O'Hara PJ, Berkner K, et al: Proc Natl Acad Sci USA 1986; 83:3101

7. Metcalf D, Begley CG, Johnson GR, et al: Blood 1986; 67:37

8. Sieff CA, Niemeyer CM, Nathan DG, et al: J Clin Invest 1987; 80:818

9. Tomonaga M, Golde DW, Gasson JC: Blood 1986; 67:31

10. Caracciolo D, Clark SC, Rovera G: Blood 1989; 73:666

11. Hoang T, Haman A, Goncalves O, et al: J Exp Med 1988; 168:463

12. Leary AG, Ikebuchi K, Hirai Y, et al: Blood 1988; 71:1759

13. Zsebo KM, Wypych J, Yuschenkoff VN, et al: Blood 1988; 71:962

14. Ikebuchi K, Clark SC, Ihle JN, et al: Proc Natl Acad Sci USA 1988; 85:3445

15. Isove NN, Shaw AR, Keller G: J Immunol 1989; 142:2332

16. Weisbart RH, Golde DW, Clark SC, Wong GG, Gasson JC: Nature 1985; 314:3651

17. Vadas MA, Nicola NA, Metcalf D: J Immunol 1983; 130:795

18. Lopez AF, Williamson J, Gamble JR, et al: J Clin Invest 1986; 78:1220

19. Fleischmann J, Golde DW, Weisbart RH, Gasson JC: Blood 1986; 68:708

20. Silberstein DS, Owen WF, Gasson JC, et al: J Immunol 1986; 137:3290

21. Grabstein KH, Urdal DL, Tushinski RJ, et al: Science 1986; 232:506

22. Weiser WY, Van Niel A, Clark SC, David JR, Remold HG: J Exp Med 1987; 166:1436

23. Sisson SD, Dinarello CA: Blood 1988; 72:1368

24. Cannistra SA, Vellenga E, Groshek P, Rambaldi A, Griffin JD: Blood 1988; 71:672

25. Horiguchi J, Warren MK, Kufe D: Blood 1987; 69:1259

26. Gasson JC, Weisbart RH, Kaufman SE, et al: Sience 1984; 226:1339

27. Arnaout MA, Wang EA, Clark SC, Sieff CA: J Clin Invest 1986; 78:597

28. Peters WP, Stuart A, Affronti ML, et al: Blood 1988; 72:1310

29. Burgess AW, Begley CG, Johnson GR, et al: Blood 1987; 69:43

30. Wong GG, Witek JS, Temple PA, et al: Science 1985; 228:810

31. Yang YC, Kovacic S, Kriz R, et al: Blood 1988; 71:958

32. Park LS, Friend D, Gillis S, Urdal DL: J Exp Med 1986; 164:251

33. Gasson JC, Kaufmann SE, Weisbart RH, Tomonaga M, Golde DW: Proc Acad Sci USA 1986; 83:669

34. DiPersio J, Billing P, Kaufman S, Eghtesady P, Williams RE, Gasson JC: J Biol Chem 1988; 263:1834

35. Kelleher C, Miyauchi J, Wong G, Clark SC, Minden MD, McCulloch EA: Blood 1987; 69:1498

36. Delwel R, Dorssers L, Touw I, Wagemaker G, Löwenberg B: Blood 1987; 70:333

37. Hoang TN, Nara N, Wong G, Clark S, Minden MD, McCulloch EA: Blood 1986; 68:313

38. Griffin JD, Young D, Herrmann F, Wiper D, Wagner K, Sabbath KD: Blood 1986; 67:1448

39. Mitjavila MT, Villeval JL, Cramer P, et al: Blood 1987; 70:965

40. Vellenga E, Young DC, Wagner K, Wiper D, Ostapovicz D, Griffin JD: Blood 1987; 69:1771

41. Delwel R, Salem M, Pellens C, et al: Blood 1988; 72:1944

42. Young DC, Wagner K, Griffin JD et al: J Clin Invest 1987; 79:100

43. Miyauchi J, Wang C, Kelleher CA, et al: J Cell Physiol 1988; 135:55

44. Park LS, Waldron PE, Friend D, et al: Blood 1989; 74:56

45. Ruff MR, Farrar WL, Pert CB: Proc Natl Acad Sci USA 1986; 83:6613

46. Berdel WE, Danhauser-Riedl S, Stainhauser G, Winton EF: Blood 1989; 73:80

47. Baldwin GC, Gasson JC, Kaufman SC, et al: Blood 1989; 73:1033

48. Anderson KC, Jones RM, Morimoto C, Leavitt P, Barnt PA: Blood 1989; 73:1915

49. Sieff CA, Niemeyer CM, Mentzer SJ, Faller DV: Blood 1988; 72: 1316

50. Yang YC, Tsai S, Wong GG, Clark SC: J Cell Physiol 1988; 134:292

51. Koeffler HP, Gasson J, Ranyard J, Souzy L, Shephard M, Munker R: Blood 1987; 70:55

52. Bagby GCJr, Dinarello CA, Wallace P, et al: J Clin Invest 1986; 78:1316

53. Kaushansky K, Lin N, Adamson JW: J Clin Invest 1988; 81:92

54. Zucali JR, Dinarello CA, Oblon DJ, et al: J Clin Invest 1986; 77:1857

55. Broudy VC, Kaushansky K, Segal GM, Harlan JM, Adamson JW: Proc Natl Acad Sci USA 1986; 83:7467

56. Munker R, Gasson J, Ogawa M, Koeffler HP: Nature 1986; 323:79

57. Miayima A, Otsu K, Schreurs J, et al: EMBO J 1986; 5:1193

58. Cantrell MA, Anderson D, Cerretti DP, et al: Proc Natl Acad Sci USA 1985; 82:6250

59. Lee F, Yokota T, Otsuka T, et al: Proc Natl Acad Sci USA 1985; 82:4360

60. Cebon J, Dempsey P, Fox R, et al: Blood 1988; 72:1340

61. Herrmann F, Schulz G, Lindemann A, et al: J Clin Oncol 1989; 7:159

62. Steward WP, Scarffe JH, Dirix LY, et al: Br J Cancer 1990; 61:749

63. Thompson JA, Lee DJ, Kidd P, et al: J Clin Oncol 1989; 7:629

64. Lieschke GJ, Maher D, Cebon J, et al: Ann Intern Med 1989; 110:357

65. Phillips N, Jacobs S, Stoller R, et al: Blood 1989; 74:26

66. Devereux S, Linch DC, Campos Costa D, et al: Lancet 1987; 2:1523

67. Baldwin GC, Gasson JC, Quan SG, et al: Proc Natl Acad Sci USA 1988; 85:2763

68. Devereux S, Porter JB, Hoyes KP, et al: Br J Haematol 1990; 74:17

69. Socinski MA, Cannistra SA, Sullivan R, et al: Blood 1988; 72:691

70. Sullivan R, Fredette JP, Socinski M, et al: Br J Haematol 1989; 71:475

71. Steward WP, Scarffe JH, Austin R, et al: Br J Cancer 1989; 59:142

72. Antman KS, Griffin JD, Elias A, et al: N Engl J Med 1988; 319:593

73. Aglietta M, Piacibello W, Sanavio F, et al: J Clin Invest 1989; 83:551

74. Broxmeyer HE, Cooper S, Williams DE, et al: Exp Hematol 1988; 16:594

75. Socinski, MA, Cannistra SA, Elias A, et al: Lancet 1988; 1:1194

76. Gianni AM, Siena S, Bregni M, et al: Lancet 1989; 2:580

77. Villeval JL, Dührsen U, Morstyn G, Metcalf D: Br J Haematol 1990; 74:36

78. Wing EJ, Magee DM, Whiteside TL, Kaplan SS, Shadduck RK: Blood 1989; 73:643

79. Evans C, Rosenfeld CS, Winkelstein A, et al: N Engl J Med 1990; 322:337

80. Gribben JG, Devereux S, Thomas NSB, et al: Lancet 1990; 335:343

81. Ho AD, Del Valle F, Engelhard M, et al: Cancer 1990; 66:423

82. Herrmann F, Schulz G, Wieser M, et al: Am J Med 1990; 88:619

83. Gianni AM, Bregni M, Siena S, et al: J Clin Oncol 1990; 8:768

84. Edmonson JH, Long HJ, Jeffries JA, et al: JNCI 1989; 81:1510

85. Brandt SJ, Peters WP, Atwater SK, et al: N Engl J Med 1988; 318:869

86. Nemunaitis J, Singer JW, Buckner CD, et al: Blood 1988; 72:834

87. Blazar BR, Kersey JH, Mc Glave PB, et al: Blood 1989; 73:849

88. Peters WO, Kurtzberg J, Kirkpatrick G, et al: Blood 1989, 74 Suppl:178a

89. Nemunaitis J, Anasetti C, Appelbaum FR, et al: Blood 1989; 74 Suppl:457a

90. Butturini A, De Souza PC, Gale RP, et al: Lancet 1988; 2:471

91. Ganser A, Ottmann OG, Erdmann H, et al: Ann Intern Med 1989; 111:887

92. Vadhan-Raj S, Buescher S, LeMaistre A, et al: Blood 1988; 72:134

93. Vadhan-Raj S, Velasquez WS, Butler JJ, et al: Am J Hematol 1990; 33:189

94. Vadhan-Raj S, Jeha SS, Buescher S, et al: Blood 1990; 75:858

95. Welte K, Zeidler C, Reiter A, et al: Blood 1990; 75:1058

96. Groopman JG, Mitsuyasu RT, Deleo MJ, et al: N Engl J Med 1987; 317:593

97. Thomssen C, Nissen C, Gratwohl A, et al: Br J Haematol 1989; 71:157

98. Vadhan-Raj S, Buescher S, Broxmeyer HE, et al: N Engl J Med 1988; 319:1628

99. Antin JH, Smith BR, Holmes W, et al: Blood 1988; 72:705

100. Nissen C, Tichelli A, Gratwohl A, et al: Blood 1988; 72:2045

101. Vadhan-Raj S, Kellagher MJ, Keating M, et al: N Engl J Med 1988; 317:1545

102. Ganser A, Völkers B, Greher J, et al: Blood 1989; 73:31

103. Herrmann F, Lindemann A, Klein H, et al: Leukemia 1989; 3:335

104. Hoelzer D, Ganser A, Ottmann OG, et al: Haematol Blood Transfusion 1990; 33:763

105. Schuster MW, Thompson JA, Larson R, et al: J Cancer Res Clin Oncol 1990; 116, Suppl:1079

106. Pluda JM, Yarchoan R, Smith PD, et al: Blood 1990; 76:463

107. Grossberg HS, Bonnem EM, Buhles WC: N Engl J Med 1989;

320:1560

108. Hittelman WN, Tigaud JD, Estey E et al: Blood 1988; 72, Suppl 1:121a

109. Vadhan-Raj S, Broxmeyer HE, Spitzer G, et al: Blood 1989; 74:1491

110. Ganser A, Ottmann OG, Schulz G, Hoelzer D: Onkologie 1989; 12:13

111. Höffken K, Overkamp F, Stirbu J, et al: Onkologie 1990; 13:33

112. Tricot GJ, Lauer RC, Appelbaum FR, Jansen J, Hoffman R: Semin Oncol 1987; 14:444

113. De Witte T, Muus P, Haanen C, et al: Behring Inst Mitt 1988; 83:301

114. Cannistra SA, Griffin JD: Third Symposium on Minimal Residual Disease in Acute Leukemia, Rotterdam 1990

115. Tafuri A, Lemoli RM, Gulati S, et al: Blood 1989; 74, Suppl 1:231a

116. Bhalla K, Birkhoffer M, Arlin Z, et al: Leukemia 1988; 2:810

117. Hiddemann W, Kiehl M, Schleyer E, et al: Blood 1989; 74, Suppl 1:230a

Discussion - THERAPEUTIC ACTIONS OF RECOMBINANT HUMAN GRANULOCYTE-MACROPHAGE COLONY-STIMULATING FACTOR (GM-CSF)

A.J.H. Gearing

The toxicities you see, with GM being more potent in toxicity than IL3 and IL3 being more than GCSF, seem to correlate with the ability to stimulate IL1 release, at least in vitro. Do you think that is reasonable?

A. Ganser

We monitored our patients checking for TNF alpha levels, IL1 levels and IL6 levels. The only cytokine we could detect was IL6 in patients developing fever. I think it is awfully difficult to detect them because where the cytokines are working is in the bone marrow or at the end organ, and one can never find IL1 or increased TNF alpha levels in the serum. I think that either their turnover in serum is too fast, as once they are released into the circulation they are taken up by the end organs, or they are not even released into the serum.

L. Gauci

One of the major concerns outside the oncology community, is the desire to use hematopoietic growth factors to allow for the increase in the amount of chemotherapy that can be given. This will certainly precipitate other serious toxicities. I think this is potentially very dangerous however it has to be tried, but should be done only by experimented research physicians. The risks are too great to permit indiscriminated usage.

A. Ganser

The hopes concerning the possibility of using greater amounts of chemotherapy were too exaggerated. For instance, one can increase the amount of adriamycin by about 50%, but then you reach toxic levels in other organs, and you have to stop there.

D. Maruhn

You mention the reduction of application of antibiotics as one of the possible advantages of administration of these drugs. Is that substantiated by any data?

A. Ganser

There are several randomized trials, both after chemotherapy and after autologous bone marrow transplantation. The multi center European trial using GM-CSF shows a shortening of the duration of neutropenia by about one week and thus the isolation time in hospital and in the laminar air flow rooms can be shortened by about one week.

J. Bigorra

It is said that interleukin 3 increases the number of platelets. Do you know if these data come in trials done in healthy volunteers or in patients, and if it was done in healthy volunteers, do you have experience in patients?

A. Ganser

The interleukin 3 phase I trial was done in patients with solid tumours but with normal bone marrow function. And we also used interleukin 3 in patients with depressed bone marrow function after previous prolonged chemotherapy or bone marrow infiltration by solid tumour cells. We have treated 10 of these patients and in 8 of them platelet counts improved. This improvement lasted for prolonged periods: even if you give it only for fifteen days, the improvement lasts for several months. Some of the patients had even been GM-CSF failures. We also used it in patients with myelodysplastic syndromes, but it is not that good in that subset of patients. Some of the patients got an improved platelet count but that was not overwhelming. Also in patients with aplastic anemia, results were quite disappointing. But I think from the data raised in patients with normal bone marrow function, that it is appropriate to combine immunosuppressive therapy and cytokine therapy in certain cases, for instance in patients with aplastic anemia.

© 1991 Elsevier Science Publishers B.V. (Biomedical Division)
The clinical pharmacology of biotechnology products.
M.M. Reidenberg, editor

POTENTIAL USE OF PEPTIDE HORMONES IN SPORT

Jordi Segura, Rafael de la Torre and Roser Badia
Department of Pharmacology and Toxicology, Institut Municipal d'Investigació Mèdica, Passeig
Marítim 25-29, 08003 Barcelona (Spain)

INTRODUCTION

The misuse of hormonal agents in sport dates back to the decade of 1960-70 where androgen-anabolic and corticoid steroids were known to be administered to athletes. These agents have continued to be used in sport although nowadays the percentage of steroids detection in sportsmen subjected to urine control is slightly declining. This is probably the result of educational and regulatory efforts but it could also be an indicator of changing trends in pharmacological abuse.

Advances in physiology and biochemistry have allowed to increase the knowledge on how the ultimate hormones are synthesized and regulated. The existence of hypophyseal and hypothalamic peptide factors is recognized together with the complicated mechanisms for the control of their blood and tissue concentrations. Human growth hormone is another hypophyseal peptide substance whose activity on the adult sportsmen has not been deeply substantiated but it is believed by some people to contribute to increase strength and endurance. The potential use of biologically active proteins and peptides has increased after the recent suggestion of sophistication in blood doping as is the misuse of erythropoietin in order to increase the number of red blood cells in athletes [1].

The availability of biological peptides has been relatively reduced until recently because the need for elaborated extraction procedures from natural sources or the complicated chemical synthetic procedures only available for small molecules. The introduction of DNA recombinant techniques is beginning to change the sources for obtaining such compounds. The misuse of such substances in healthy people will generate unknown problems to the delicate hormonal equilibrium of the human body. Taken all this into account the International Olympic Committee has included the peptidic hormones and analogs either from synthetic or recombinant origin among the list of forbidden substances in sport. This chapter presents a report about those hormones that may be potentially misused. Those that are of non-recombinant origin nowadays may become it in a few time.

ACTION ON GLUCOCORTICOID METABOLISM

Adrenocorticotropic hormone ACTH

Cortisol is a glucocorticoid synthesized in the adrenals which allows the body to face uncommon and stressful situations by affecting a number of reactions of intermediate metabolism. The development of physical exercise is one of the stimulus that results in an increase of serum cortisol. Psychic pressure is known also as a factor resulting in increased cortisol levels. Recent research [2] has demonstrated that high level sport competitions makes the athlete to be confronted to this double mechanism in cortisol secretion. The results reported in Figure 1 demonstrated that only the competition, but not a normal training nor an exhaustion treadmill test generates the

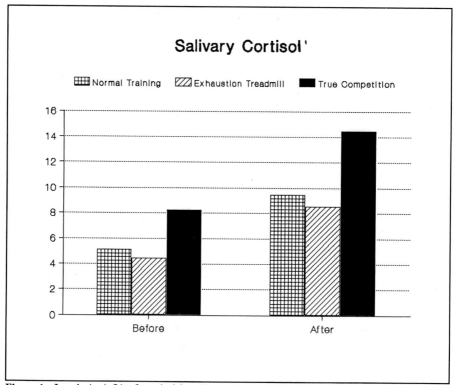

Figure 1.- Levels (ng/mL) of cortisol in saliva before and after three different situations in top level swimmers

high cortisol response. Thus the possibility exists of artificially increasing cortisol levels, specially in pre-competition situations, to artificially habituate the athlete to an actual competition setting.

Synthesis of cortisol by the adrenal cortex is stimulated by ACTH, a polypeptide hormone secreted by cells of the anterior pituitary gland. Thus the exogenous administration of ACTH might be used to further stimulate cortisol secretion overriding feed-back inhibition or just generating an increase of cortisol over normal values. In either case it would be a non-physiological use of corticotropic hormone.

Human ACTH is composed of 30 aminoacids with a molecular weight of 4500. Actual sources of ACTH include extraction from animal pituitaries by chromatographic or electrophoretic techniques and the chemical synthesis of the entire human ACTH polypeptide. However, not all the aminoacid structure is needed for biological activity. The synthetic analog of the ACTH(1-24) is named Cosyntropin and is currently used by intramuscular or intravenous injection of about 0.25 mg for diagnostic purposes. ACTH(1-17) is also available.

Corticotropin-releasing hormone (CRH)

In addition to the direct negative feed-back mechanism, the secretion of ACTH is regulated by the CRH formed in the hypothalamus and released to the hypothalamic portal venous system, thus reaching the pituitary corticotropic cells. The structure of this factor (a 41 aminoacids polypeptide) was known in 1981. It is presently used for diagnostic purposes but it could be misused to artificially trigger the release of ACTH and consequently cortisol. It is obtained by isolation and purification from bovine hypothalami. Its synthesis by recombinant techniques might be developed in the near future.

ACTION ON ANDROGEN SECRETION

Gonadotropins

Luteinizing hormone (LH) and follicle stimulating hormone (FSH) are glycoproteins synthesized by the gonadotropes of the anterior pituitary gland. Human chorionic gonadotropin (HCG), also a glycoprotein, is synthesized by the placenta. Each glycoprotein is made of two dissimilar, glycosylated polypeptide chains, alpha and beta subunits, linked by hydrogen bonds. The alpha subunit is essentially common to LH, FSH, HCG and TSH (Thyroid-Stimulating hormone) and comprises 92 aminoacids . The beta subunit identifies the appropriate target tissue and confers their biological activity. The beta subunit has 121 aminoacids for LH,118 aminoacids for FSH and 145 aminoacids for HCG [3]. HCG beta-chain have a structural similarity with the beta-chain of LH; 97 aminoacids of HCG beta-chain are identical to those of beta-LH [4]. HCG has higher sugar content than LH, and has a longer half-life. The molecular weight of gonadotropins depends on carbohydrate composition and is about 30000 for LH, 36000 for FSH and 65000 for HCG.

HCG is not only produced by placenta. Trophoblastic tissues which can be found in hydatidiform mole, choriocarcinomas and germ cell tumors of the testes also produce HCG. It is also found in solid tumors of the ovaries, uterus, breast, lung, pancreas and stomach [5] [6]. HCG can not be considered a foreign substance in the male, but the amounts found in normal men are minute. A recent study [7] concludes that HCG is produced in a pulsatile fashion, probably by the pituitary, in all normal adults. The amount found in serum samples from normal men averaged 8.9 pg/mL (range 3.0-160 pg/mL) (biologic potency 13.450 IU/mg).

Due to the similarity between HCG and LH beta chain, HCG injected in a male may bind to the LH receptors in the Leydig cell and stimulates the biosynthesis and excretion of testicular steroids, increasing testosterone and epitestosterone level. Brooks et al. have reported [8] that a single injection of testosterone heptanoate (Primoteston), 250 mg, produced a 3-5 fold increase in the plasma testosterone concentration but it declined fairly rapidly. Intramuscular injection of 5000 IU HCG, for four days, increases 2 to 3 fold testosterone level at the end of this period. With administration of HCG the maximum concentrations of testosterone in plasma were not as high as those obtained after Primoteston, although the rise was quite well sustained. This study also showed that HCG can be found in the urine at concentration above the cut-off limit for as long as the concentration of testosterone in the plasma is raised above the basal concentration.

Gonadotropin releasing hormone

(GnRH), also known as luteinizing hormone releasing hormone (LHRH), is a decapeptide that has a molecular weigh of 1182 and hat is synthesized by neurons in the brain. It is released from nerve terminals and is carried to the anterior pituitary by the hypophyseal portal vessels, where it stimulates the production and release of both LH and FSH. GnRH half-life is only of five minutes because is quickly metabolized by endopeptidases.

GnRH Analogs differ from the natural hormone in some peptide positions [9] [10]. They have more affinity for the hypophysis receptors and more stability in front of enzymatic degradation. Agonists analogs release more gonadotropins and in a more lasting time than natural hormone. Depending on the dose and way of administration of GnRH or its analogs, a stimulation or an inhibition in the pituitary is produced [11]. A pulsatile exogenous administration of GnRH increases gonadotropin production, while a continuous administration of GnRH inhibits gonadotropin release.

Table 1.- <u>Some Gonadotropin-related pharmaceutical Products</u>

Chorionic Gonadotropin (from the urine of pregnant women)

Menotropin (LSH and FSH from urine of postmenopausal women)

Gonadorelin (synthetic Gonadotrophin releasing hormone)

Gonadotropin releasing hormone Analogs
    Buserelin Acetate
    Goserelin Acetate
    Leuprorelin Acetate
    Nafarelin Acetate
    Triptorelin

HCG components in urine

Several authors have identified different immunoreactive HCG components in urine [5]: Intact HCG, beta-subunit, alpha-subunit and fragments of the beta-subunit. There is a variability in sialylation and glycosylation such that the same polypeptide species may appear in several different molecular weight fractions.

Immunological methods of detection

Several immunological methods have been developed for the detection and quantitation of gonadotropins. Due to its structural similarity of alpha-subunit, exclusively antisera against the LH, FSH, HCG or TSH beta-subunit gives specificity to the method. The homology between LH and HCG beta-chain can result in cross-reactivity. In serum, proteins that bind to gonadotropins specific antibodies may cause aberrant test results [5].

The use of monoclonal antibodies has the advantage of very high specificity. However, they may miss some molecules with either slight modifications, denaturation, or steric hinderance at the specific binding site. The result is a diminution of sensitivity. Polyclonal antibodies react with more sites on the molecule, which should provide increased sensitivity and could result in detection of more types of fragments and metabolites. A disadvantage of the polyclonal systems may be a small increase in their ability to bind substances with structural homologies. So, test with polyclonal antibodies are less specific than test with monoclonal antibodies.

A great variety of immunoassays which used different labels such as radioisotopes, fluorescent molecules, enzymes or others are currently available. Also, there are immunoassays carried out in solution or in gel, with liquid phase or solid-phase reagents and so on [4]. A comparative study carried out by Brooks et al. [8] between various disposable commercial kits: Serono MAIA clone Immunoradiometric assay (IRMA), Serono Serozyme Enzymoimmunoassay (EIA) and Boehringer Enzyme linked immunoassay ELISA, showed a good performance of all of them. The specificity is ensured by three high affinity monoclonal antibodies to both the intact HCG molecule and to the beta subunit. The cut-off limit has been set far in excess of the level found in male urines from subjects who have not been given HCG.

Stenman et al. [12] had described a combination of chromatography with immunoassay method. It could be used in doping control [8]. The method uses gel chromatography on a TSK HW-50 (S) column and radioimmunoassay of the fractions using LH and HCG antisera.

GROWTH HORMONE AND ITS RELEASING FACTOR

Chemistry and Physiology

Growth Hormone (GH)

GH is an anterior pituitary hormone that has a molecular weight of 21000 and comprises 190 aminoacids in a simple chain with two intrachain disulphide bridges. GH acts throughout the body to stimulate the longitudinal growth of bones and to regulate glucose, amino acid, and fatty acid metabolism. Many of the actions of GH are mediated by other hormones released from the liver called somatomedines, or insulin-like growth factors [13].

Growth hormone release is controlled by two hypothalamic factors: the growth hormone releasing factor (GHRH), which stimulates GH secretion, and somatostatin, which inhibits GH secretion. The GH-RH molecule consist of 44 aminoacids. Other isolated shorter fragments 1-37 and 1-40 are also biologically active. A number of metabolic factors influence the secretion of GH. Increased plasma levels of substances such as glucose, non-sterified fatty acids and ketone bodies inhibit GH release, whereas hypoglycemia stimulates GH secretion. Consistent with these observations is the elevation of GH secretion during strenuous exercise.

Growth hormone in use in therapeutics must be from human origin because other animal sources are biologically inactive. The source of GH used in clinical situations in the past has been cadaver pituitaries. Severe controls are needed to reduce bacterial and viral contamination of preparations of GH from human pituitaries. So, in some countries GH preparations from recombinant DNA techniques are preferred. Growth hormone preparations from recombinant DNA methodologies are produced after the insertion of the GH messenger RNA gene in a plasmid and its expression either in procaryotes (E. coli, Somatrem) or mammalian cell lines [14].

Therapeutic use of GH-RH is mainly restricted to dwarfism of hypothalamic origin. Little is known about side effects in normal adults and mimetically we associate those related with clinical situations where there is an hypersecretion of GH. Side effects of hypersecretion of GH in adults are those associated with acromegaly. The most prominent are myopathy, peripheral neuropathy, coronary artery disease and cardiomyophaty [15].

Table 2.- Growth Hormone/Growth Hormone-Releasing Hormone Preparations

1.-Growth Hormone Preparations [16] [17]

    Somatropin Refers to preparations of GH with the same sequence of the natural human GH either pituitary-derived or synthesized by means of DNA recombinant techniques.

    Somatrem Refers to a synthetic methionyl analogue of GH.

2.-Growth Hormone-Releasing Hormone Preparations [16]

    Somatorelin Refers to preparations of GH-RH with the same sequence of the natural human GH-RH either hypothalamic-derived or synthesized by means of DNA recombinant techniques. Usually the DNA recombinant preparation consist on the 1-40 aminoacid sequence of the naturally occurring GH-RH.

    Sermorelin Refers to a synthetic version of GH-RH with the first 29 aminoacids of its natural sequence.

Pharmacokinetics

After the administration of recombinant GH by the intravenous route the mean half-life is 9 minutes [18] although some reports presents longer half-lives estimations (around 25 minutes). Maximal effects are reached between 3 and 5 hours depending on the administration route (3 hours IM, 4-5 hours S.C.) [18] [19]. When interpreting these pharmacokinetic data, we must take into account that some effects of GH are mediated by hepatic somatomedines (half life between 3 and 4 hours) and that the effects last for longer periods of time. Reports show that recombinant GH is safe and well tolerated by the intravenous, intramuscular and subcutaneous routes in short term periods [20]. The preparations somatropin and somatrem are bioequivalent [21].

Most of the pharmacokinetic data of GH-RH preparations have been evaluated on the basis of they maximal effects on GH secretion. Different administration routes have been assayed including the intranasal, intravenous and intramuscular. Maximal effects appear between 15 and 40 minutes after the administration irrespectively of using sermorelin or somatorelin [22] [23] [24]. Both preparations are bioequivalent [24].

Assessment of GH and GH-RH in body fluids

1.-Assay

Radioimmunoassay (RIA) and other non-isotopic immunological techniques have replaced bioassays and are routinely used in the assessment of GH concentrations in patients sera and for physiological studies on GH secretion [25] [26] [27] [28] [29] [30].

Important differences have been found between techniques using monoclonal and polyclonal antibodies. Most of these techniques are standardized with GH from pituitary origin but interestingly there was more than a fivefold variation in cross reactivities against dimeric GH from DNA recombinant techniques.

If natural sequences of GH and GH-RH are administered no assay will distinguish between endogenous and exogenous materials because antigenically they are indistinguishable. Short after the administration of these substances levels of GH and GH-RH may be unnaturally high in plasma. At present is not known what kind of changes we can expect in urine. If analogs of GH or GH-RH are administered, there is a chance of developing immunoassays to detect them. Other possibilities are the detection, with very sensitive immunoassays, of antibodies against these substances.

Alternatively it has been proposed to use assays of somatomedine C and the insulin growth factor 1 as a measure of the degree of exposure to biologically active GH in humans [15],[31].

Some techniques using mass spectrometry coupled to high performance liquid chromatography have been used successfully for the characterization of natural and biosynthetic GH, and are promising for a future application in biological fluids [32].

2.-Urinary excretion of GH

Growth hormone and the growth hormone binding protein (GH-BP) can be detected in urine using current immunological analytical techniques [33] [34]. The detection of GH-RH in urine has not been reported yet.

Some aspects are still pending before starting urinary controls of GH. In one hand normal values for normal people -but most specifically in highly trained/physically stressed athletes- must be validated. In the other hand we must establish which compounds GH like are being detected in urine. Detection of GH in urine is based on cross reactivities with different antibodies raised against plasma products. Some preliminary results show that compounds detected in urine are not the parent compounds GH and GH-BP but these compounds metabolized to a certain extent [35].

ERYTHROPOIETIN

One way of affecting the blood components is the use of hematopoietic growth factors. Many of them are now available in large quantities through recombinant DNA technology [36],[37]. Erythropoietin was the first factor studied and is the only clinically available at present. It acts on the mature red cell precursors to induce their proliferation and differentiation. The concurrent use of erythropoietin with other growth factors such as Interleukin-3 (IL-3) or the Growth factor for granulocytes / macrophages (GM-CSF) might be used in the future to stimulate even those other early developing cell precursors.

The Product

Erythropoietin is a glycoprotein of about 30400 amu whose aminoacid chain has 165 residues (molecular weight 18224 amu). The protein contains 2 disulphide bonds (positions 7-161 and 29-33), 3 asparagine N-glycosylations (positions 24,38 and 83) and 1 serine O-glycosylation (position 126). Glycosylation is not, however, necessary for biological activity but protects the protein molecule from a rapid degradation. Characterization of EPO structure followed the observation of its presence in urine [38] and its subsequents isolation [39].

Present availability of rHuEPO is consequence of cloning and expressing the responsible gen (7q 11-22). Initial synthesis using E.coli cells have been substituted by mammal cells (Chinese Hamster Ovary) in order to afford complete glycosylation. Only a few pharmaceutical companies have developed the recombinant EPO product. Under license it is being approved in a number of countries.

Physiology

Erythropoietin is produced mainly in the kidney (about 90%). The production of erythropoietin is regulated by the concentration of oxygen in the blood. EPO stimulates the final differentiation of committed erythroid progenitors, increases hemoglobin synthesis and promotes the release of marrow reticulocytes into the circulation.

Main clinical indication of rHuEPO is for those anaemias due to low EPO availability such is the chronic renal failure [40]. Experiences in other kind of anaemias of inflammatory or neoplastic origin is being accumulated. It has also been suggested to use EPO for increasing the number of erythrocytes for those patients to be bled and to be further autotransfused. A similar misuse in sport would generate a double incorrect process (EPO+blood doping).

Hypertension is the major problem in EPO treated patients. In fact, increase in blood pressure has also been observed in healthy people treated with EPO [41]. Some other side effects described in treated patients are seizures and increments in serum potassium and urea and in the thrombocyte count (clotting may be at increased risk). An unexplained influenza-like syndrome related to EPO administration has rarely appeared.

Pharmacokinetics

A single dose application to patients by intravenous route resulted in a serum half-life of 5.4 hours (dose 50 UI/Kg) or 7.6 hours (dose 150 UI/Kg) [42]. Other reports about serum EPO half-life in patients range from 4 to 11 hours [40] [43]. The concentration of serum EPO after subcutaneous administration is many times lower (5-10%) with a concentration peak obtainable between 12 and 24 hours [40] [44].

In spite of the pharmacokinetics described, a dose of 50 IU/Kg 3 times weekly is usually used and further titration for each patient is done according to the respective response. Higher doses may reach around 450 IU/Kg weekly. Once a target hemoglobin level has been reached reductions in dose frequency to twice or even once weekly are possible [40].

Reports from EPO administration to healthy people as it would be the situation in athletes are scarce in the literature. A recent report [45] evaluates the pharmacokinetics after both IV or SC administration in six healthy people. Volume of distribution and half life after iv route were 76 ml/kg and 4.5 hours respectively. Clearance was 64 ml/kg which is slightly higher than in patients, probably because a more efficient urinary excretion in healthy people. However, recent data on renal excretion of rHuEPO in healthy subjects [43] indicate that total urinary excretion represented less than 5% of the dose at all dose levels. Bioavailability by subcutaneous administration was 36% over a period of 72 hours and half-life was estimated to be around 46 hours. Subcutaneous administration originates smaller fluctuations and a sustained increase over baseline levels.

When there is no EPO administration, basal levels of EPO in serum in healthy people are about 10-50 mU/ml. Under treatment with rHuEPO there appear to be a proportional relationship between serum concentrations and single dose administrations. Levels in the range of 20 mU/ml to 30 U/ml have been reported [43] after intravenous dosages (10-1000 IU/Kg). Levels substantially higher than normal (up to 5-30 U/ml) are found also in acute and aplastic leukemia and in myelodisplastic disorders [46].

Effects on exercise related variables

a) Patients

Since initial trials with hemodialysis patients [47] it was clear that in addition to correct anaemia, exercise capacity and perhaps cardiac function could be improved by EPO treatment. A multicenter,placebo controlled, clinical trial with hemodialysis patients where functional capacity was measured on a treadmill [48] lead to the conclusion that EPO causes a marked statistically and clinically significant improvement towards fatigue and physical symptoms. EPO appeared to cause an improvement in exercise capacity, but the magnitude of the change was not as great.

b) Athletes and healthy people

With regard to EPO in healthy adults there is little clinical information available. Main data come from a study performed at the Karolinska Institute [41] (to be published) where young gymnastics were treated with either rHuEPO, autologous blood or placebo [49]. In regards to the reponse elicited by EPO [41], an increment of oxygen capacity and endurance to treadmill running was observed. Hemoglobin content and the hematocrit was also increased but not the total blood volume. Maximum aerobic power returned to normal after 2 weeks of cessation of EPO treatment.

Detection of rHuEPO

Theoretically there appear to be different ways for trying to identify the misuse of rHuEPO in sport although none of them has been successfully applied to date. Some of them are listed below:

a) Speculations about slight differences between the recombinant product and the natural hormone are the base for the suggestion of finding specific antibodies to the recombinant product as an index of previous use. So far no antibodies to the artificial product have been found in patients receiving the product [50].

b) Detection of specific modifications of the red cell biochemistry and morphology that may follow EPO treatment [51].

c) To request regulatory bodies to ask companies involved in producing rHuEPO to add an internal marker compound to the marketed products [52].

d) Detection of <u>urinary excretion of rHuEPO</u>. Obviously some critical quantitative cut-off point after population studies would be needed to distinguish, if possible, excretion of exogenous rHuEPO from normal endogenous EPO excretion because no biochemical distinction is expected.

e) <u>Determination of the blood EPO concentrations</u>. Higher than normal values may be found in rHuEPO users, but only for a few hours after intravenous, and for a few days after subcutaneous administration [51]. Unambiguous cut-off distinction would be needed as stated above. Traditional research methods to detect EPO in biological samples are based in bioassays. The in vivo bioassays [53] are the most reliable assays for biologic EPO activity and they should be used as reference to compare immunoreactivity and bio-activity of any other methodology [54]. In vitro bioassays [53] appear to have too many pitfalls (i.e. false positive reactions) that render them inappropriate for their application [54].

The developments of immunological methods will allow routine determination of blood EPO in the near future. The limit of detection for some RIA assays [45] [46] [55] [56] range from 2-10 mU/ml with coefficients of variation around 6-7%. The alternative use of non isotope labeling is used in recently developed ELISA assays [57] [50] with similar technical characteristics.

MISUSE IN SPORT

<u>Gonadotropins</u>

The abuse of testosterone may be detected by measuring testosterone to epitestosterone ratio (T/ET) with the established GC/MS techniques [58]. But this ratio remains normal after HCG abuse. In 1986, 42 samples were screened using T/LH ratio [8]. Two of these samples, that come from male, were found to have low T/LH ratio suggesting that HCG might have been administered. Assayed by the MAIA clone method for HCG, these two samples were found to have values of 242 and 154 IU/L, while the normal range for this method is 1.2-3.9 IU/L.

In 1987, 740 male sport urines were tested by this method and 21 positive urines having HCG concentrations in the range 30-119 IU/L were found [8]. In 1988 one HCG positive urine had been reported and 3 cases have been reported in 1989.

A recent study [59] suggests that HCG could be used as a masking agent after testosterone abuse, if only urinary T/ET ratio is used as a doping criterion. This ratio remains normal in individuals injected with testosterone coupled with HCG. T/LH ratio appears effective as an additional marker for the detection of testosterone coupled with HCG doping.

Since now no case of GnRH or its analogs has been reported.

<u>Growth hormone</u>

The ergogenic properties of growth hormone have been the basis of its use in sports. Even if there are some reports that show an increase in muscular development after GH administration, this muscular hypertrophy is not accompanied by an increase in strength. Despite the fact that any scientific report has been able to demonstrate an increase in athletic performance its consumption is becoming widespread within athletes that believe that with the current doping control procedures it cannot be detected.There are anecdotal reports of misuse of GH in sports. Some cases of athletes with early signs of acromegaly have been detected [15] [60].

There are some reports that suggest that GH is used in very short periods of time at the end of a cycle of treatment with anabolic steroids. A synergistic effect between anabolic steroids and GH and/or GHRH has been proposed.

A new possible form of "GH misuse" is to increase indirectly GH secretion by the ingestion of some drugs and aminoacids [60].

Erythropoietin

The misuse of erythropoietin has not been substantially reported although speculations are abundant among the non technical press [61]. Rumors have been mainly focused on distance runners and cyclists. In fact it appears to have been considered, inter alia, as a potential explanation for some fatalities studied by the Royal Dutch Cycling Federation. Because the ergogenic potential of its misuse, similar to blood doping, some sport federations (i.e. cross country skying) are considering its control by blood analysis.

Final remarks on detection

The control of non-physiological use of peptide hormones in sport is additionally complicated by different factors:

a) They have a very short plasma half-life and therefore they are excreted, if any, in very small quantities in urine (the normal fluids for antidoping analysis). Even the exact chemical structures of the products excreted in urine, coming from metabolism in the body, are not usually known.

b) The majority of analytical techniques are based on radioimmunoassay which means a relative non-specificity of the antibodies used and the possibility of cross-reactivity with endogenous compounds. Different specificity of antibodies or mixtures of them used by different techniques is additionally complicated by their monoclonal or polyclonal origin. Availability of Internationally acceptable Reference Materials in order to standardize results is absent for some of the hormones. Concentrations reported by using a given methodology may not usually be compared to other assays for the same hormone.

c) Even when specifically and sensitively detected it is not easy to establish criteria to distinguish an exogenous administration from a normal endogenous concentration.

d) Finally, the total confirmation of the structure of the peptide or protein detected should ideally be done by mass spectrometry. Only recently new techniques (i.e.electrospray, fast atom bombardment) are becoming potentially available to be used as confirmation assays.

Acknowledgements

The authors especially appreciate the suggestions and informations received from Prof. A.H.Beckett (London). Additional informations received from Prof. F.Conconi (Ferrara), Prof. C.Bauer (Zurich) and Dr. Wieczorek (Mannheim) are also deeply acknowledged.

REFERENCES

1. Wadler G.I.,Hainline B. (1989) in Drugs and the Athlete,   F.A.Davis Co., Philadelphia 172-177

2. Aubets J., Torres M., de la Torre R., Camí J., Segura J.   (1989), VIIth European Congress in Sports Medicine, Barcelona

3. Stockell A. (1989) Multiple Forms of Pituitary and Placental Gonadotropins. Oxf Rev Reprod Biol 11:147-177.

4. Kiburis J. (1989) IInd I.A.F.World Symposium in Sport. Monte Carlo.

5. Rundell C.A. (1986) Journal of Clinical Immunoassay 9(2):89-95.

6. Hussa R.O. (1981) Ligand Rev 3:6-44

7. Odell W.D., Griffin J. (1987) The New England Journal of Medicine 317(27):1688-1691

8. Brooks R.V., Collyer S.P., Kicman A.T., Southan G.J., Wheeler M.A. (1989) IInd I.A.F. World Symposium on Doping in Sport. Monte Carlo.

9. Schally A.U., Arimura A. Coy D.H. (1981) Vitam Horm 38:257

10. Delfosse M. (1989) J Pharm Belg 44(2):137-142

11. Calaf J., Guinot M., Domingo N., (1988) JANO 34(815):1689-1693

12. Stenman H., Alfthan H., Ranta T., Vartiainen E., Jalkanen J., Seppala M. (1987) J Clin Endocrin Metabol 64:730-736

13. Brown B.L., Forsling M.L., Gould R.P. (1982) In: O'Riordan J.L.H., Malan P.G., Gould R.P. eds. Essentials of Endocrinology. London: Blackwell Sci. Publ.,25-57.

14. Haynes S.P. (1987) In: Belloti P., Benzi G., Ljungqvist A. eds. Official Proceedings. Ist I.A.F. World Symposium on Doping in Sport. Florence,175-92.

15. Wadler G.I.,Hainline B. (1989) in Drugs and the Athlete,   F.A.Davis Co., Philadelphia 70-74

16. Martindale, The Extra Pharmacopeia (1989) 29th Edition. The Pharmaceutical Press,London 1148-51

17. Catalogo de Especialidades Farmaceuticas (1990) Consejo General de Colegios de Farmaceuticos eds. Madrid

18. Hindmarsh P.C., Matthews D.R., Brain C.E., Pringle P.J., di Silvio L., Kurtz A.B., Brook C.G. (1989) 30:443-50

19. Ho K.Y., Weissberger A.J., Stuart M.C., Day R.O., Lazarus L. Clin Endocrinol 1989;30:335-45.

20. Sietnieks A., Wilton P. III World Conference on Clinical Pharmacology & Therapeutics. Stockholm,1986:333.

21. Wilton P., Widlund L., Guilbaud O. (1987) Acta Paediatr Scand Suppl 337:118-21

22. Albin H., Auziere J., Bedjaoui A., Angibeau M., Morre M. (1986) III World Conference on Clinical Pharmacology and Thearpeutics. Stockholm, 334.

23. Colle M., Frangin G., Jacob C., Angibeau R., Coude F.X. (1986) III World Conference on Clinical Pharmacology and Toxicology. Stockholm, 314.

24. Vance M.L., Evans W.S., Kaiser D.L., Burke R.L., Rivier J., Vale W., Thorner M.O. 1986 Clin Pharmacol Ther 40:627-33.

25. Evans A.J., Wood P.J. (1989) Ann Clin Biochem 26:353-7

26. Hashida S., Ishikawa E., Mohri Z., Nakanishi T., Noguchi H., Murakami Y. (1988) Endocrinol Jpn 35:171-80

27. Hourd P., Edwards R. (1989) J Endocrinol 121:167-75

28. Brostedt P., Luthman M., Wide L., Werner S., Roos P. (1990) Acta Endocrinol 122:241-8

29. Tomita H., Ogawa M., Kamijo T., Mori O., Ishikawa E., Mohri Z., Murakami Y. (1989) Acta Endocrinol 121:513-9

30. Bowsher R.R., Apathy J.M., Ferguson A.L., Riggin R.M., Henry D.P. (1990) Clin Chem 136:362-6

31. Hizuka N., Takano K., Asakawa K., Sukegawa., Horikawa R., Yoshizawa Y., Saito S., Shizume K. (1988) Acta Paediatr Scand Suppl 347:127-33

32. Nakazawa H. (1988) Chem Pharm Bull 36:988-93

33. Hattori N., Shimatsu A., Kato Y., Imura H. (1990) Kidney Int. 37:951-954

34. Hattori N., Shimatsu A., Kato Y., Koshiyama H., Ishikawa Y., Tanoh T., Assadian H., Imura H. (1989) Acta Endocrinol 121:533-7

35. Hashida S., Ishikawa E., Mohri Z., Nakanishi T., Noguchi H., Murakami Y. (1988) Endocrinol Jpn 35:171-80

36. Cheson B.D. (1990) Drug News Perspect. 3:154-160

37. Whetton A.D. (1990) TIPS 11:285-289

38. Finne P.H. (1965) Br Med J 697-699

39. Miyake T., Kung C., Goldwasser E. (1977) J Biol Chem 252:5558-64

40. Winearls C.G. (1989) Drugs 38:342-345

41. B.Ekblom (1990) IOC Medical Commission Meeting, Laussanne 7-8 April

42. Nielsen O.J. (1990) Pharmacol Toxicol 66:83-86

43. Flaharty K.K.,Caro J.,Ersley A.,Whalen J.J.,Morris E.M.,Bjornsson T.D.,Vlasses P.H. (1990) Clin Pharmacol Ther 47:557-563

44. Lui S.F., Chung W.W., Leung C.B., Chan K., Lai K.N. (1990) Clin Nephrol 33:47-51

45. Salmonson T.,Danielson B.G.,Wikstrom B. (1990) Br J Clin Pharmacol 29:709-713

46. Hamazaki H. et al. (1990) Kinki University School of Medicine 23:975-985

47. Esbach J., Egrie J., Downing M., Browne J., Adamson J. (1987) New Engl J Med 316:73-78

48. Mandin H. (1989) IInd I.A.F. World Symposium on Doping in Sport, Monte Carlo

49. Wieczoreck L. (1990) personal communication

50. Kientsch-Engel R.,Hallermayer K., Dessauer A. (1989) Contrib. Nephrol. 76:100-105

51. Conconi F. (1990) IOC Medical Commission Meeting, Laussanne 7-8 April

52. Subcommission Doping and Biochemistry of Sport of IOC Medical Commission (1989) Position paper on Erythropoietin

53. Kurtz A., Eckardt K-U. (1989) Nephron 51(suppl 1):11-14

54. Bauer C. (1990) personal communication

55. Eckardt K.-U., Kurtz A., Hirtz P., Scigalla P., Wieczorek L., Bauer C. (1988) Klinische Wochenschrift 66:241-245

56. Matsubara K.,Yoshimura T., Kamachi S., Fukushima M., Hino M.,Morli H. (1989) Clin Chim Acta 185:177-184

57. EPO-ELISA Test (Medac Diagnostika)

58. Massé R., Ayotte C., Dugal R. (1989) J Chromatogr 489:23-50

59. Ueki M. (1990) personal communication

60. Macyntire J.G. Growth Hormone and Athletes. Sports Med 1987;4:129-42.

61. Walker R., Brown A. (1988) Calgary Herald, Feb 17
    Jereski L. (1989) Business Week, Dec 11
    Temple C. (1990) The Sunday Times, Jan 7
    Unknown (1990) El País, Jan 17
    Specter M. (1990) Herald Tribune, Apr 3
    Highfield R. (1990) Daily Telegraph, Apr 12

Discussion - POTENTIAL USE OF PEPTIDE HORMONES IN SPORT

M.M. Reidenberg

A couple of questions. First, is there any evidence that gonadotropin used actually improves performance. From what I have read the modest doses of androgens make little difference. The people who bulk up take enormous doses of androgen and that is most unlikely to be able to come endogenously no matter how much gonadotropin you give the testis. Secondly, I can see that cortisol given in an acute situation could theoretically help, but I think there is no data to suggest it. If one gave ACTH chronically during training its antianabolic effect ought to be detrimental to performance rather than beneficial. And so I am puzzled why these pituitary hormones would be used at all for athletes; they ought to be detrimental if anything.

J. Segura

The appearance of some compounds as banned substances is not always substantiated by scientific studies. Some of the substances are there because of logical reasons, but not just because they have been proved to be useful to athletes. Releasing factors are included probably to complete the list of hormones and main factors affecting them. I agree that ACTH administration without a very special knowledge of its effects could even be detrimental to the athlete.

M.M. Reidenberg

But then if there is reason to suspect it will be detrimental why does the Olympic Committee add to the burden of the analytical chemist by banning it and forcing you to develop methods to measure compounds that don't matter whether the athlete takes or not, so far as the potential performance is concerned.

L. Gauci

Most of the drugs that you are talking about have to be given by injection. How many injection needle marks should we accept from a normal healthy athlete?

A.J.H. Gearing

Can you also indicate what the major source of these drugs is? Are they normal pharmaceuticals that have been purloined from pharmacies or are they research grade materials? And where is the normal source?

J. Segura

This is something that really puzzles. The majority of these sophisticated drugs are released in a very controlled environment. In many countries there are even regulatory bodies that control the distribution of the drug on a patient's name basis. So, it seems difficult that these drugs could be diverted into the black market, but obviously this is the case for some of them.

W. Aulitzky

The use of androgens has two purposes. First of all, the increase of muscle weight which is important during the training period. Second, and this is a much more important reason, it causes aggressive behaviour. We know that intensive training results in a decrease of pituitary hormone release and to cope with that people are using LHRH or ACTH to overcome those periods of fatigue and bad performance. As a matter of fact we are in the stage where we may loose sports as one of the major educational tools because some sports are totally dominated by pharmacology, biochemistry or medicine.

D.C. Brater

Concerning the use of internal markers, I would think that a relatively safe way to incorporate internal markers would be to put into some of the synthetic steps some stable isotopes, which could be readily detected by mass spectrometry. I don't know what the regulatory implications of that would be, but I personally wouldn't be bothered by taking something that had a stable isotope in it. In fact some of your detection methods might be made very easily; if it was a stable carbon you could even do breath analysis, and you wouldn't have to wait in the locker room until they could generate a urine sample.

J. Segura

The incorporation of stable isotopes is one of the markers that have been suggested. But it is really difficult to state if the small percentage of the drug that will be misused justifies that inclusion into the product. There are probably alternative ways to cope with that problem of hormone misuse in sport other than the incorporation of these expensive and not totally physiological markers to a molecule released for a wider clinical use.

J. Mous

Talking about growth hormone, if I remember correctly, many of the effects of growth hormone are mediated through the induction of insulin-like growth factors. Should these not be taken up into the list of forbidden products?

J. Segura

Really I don't know. It would depend on the availability of the products. Do they exist as a product that can be administered? In fact, it has been suggested that we should monitor insulin-like growth factors as indicators of human growth hormone abuse.

J. Bigorra

I don't know whether it makes sense but I wonder whether some very simple biological parameters such as hematocrit or reticulocyte count could provide a clue for the pre-screening of EPO. I mean not as detection method or as a confirmatory measure, but only to decide whether further testing is indicated or not.

J. Segura

Well, for instance, to detect blood doping there is already a strategy that, I would say, is quite similar to what you suggest. We check the blood levels of EPO, the haemoglobin level and the content of iron. And these three parameters together can indicate probably with 60% exactitude that some autologous or heterologous blood doping has been done. But the problem is that in doping control, because there is some punishment later on, we can not cope with a probability. We have to resort to very specific methodology that can allow us to claim with more or less 100% security that this or that has been consumed or not.

BIOTECHNOLOGY AND SOCIETY

© 1991 Elsevier Science Publishers B.V. (Biomedical Division)
The clinical pharmacology of biotechnology products.
M.M. Reidenberg, editor

COSTS OF DISCOVERING, DEVELOPING, MANUFACTURING AND MARKETING BIOTECHNOLOGY PRODUCTS

WILLIAM M. WARDELL

Boehringer Ingelheim Pharmaceuticals, Inc., 90 East Ridge, P.O. Box 368, Ridgefield, CT 06877

INTRODUCTION

The costs of biotechnology products are important to both the producer and the patient. From the producer's viewpoint, development costs are a major determinant of return on investment, which is an important factor in deciding whether a pharmaceutical company will attempt to translate a biological discovery into a therapeutic drug.

Costs also have a direct bearing on the pharmaceutical industry: rising R&D costs are one of the main reasons for the growing trend of mergers and consolidations. Costs influence patterns of international resource allocation, and hence international competitiveness, and have become an issue in their own right in the seemingly continuous policy debates on the pharmaceutical industry.

In the case of biotechnology, costs are of special importance for several reasons:

o   The nature of the industry (in particular the predominant role of small startup companies),

o   The inverse relationship between specificity of therapy and size of the population eligible for treatment: as therapy becomes more specific, the potential number of patients may diminish,

o   The need for special production facilities and
    quality control,

o   The relatively high ingredient cost of the final
    product,

o   The attitude of payers--either the patient or third
    parties.

There has been no comprehensive study of the costs of
developing biotechnology drugs, although the literature
gives us references to certain parts of the process and one
overall example (IL-2). Furthermore, the question is
becoming somewhat blurred as the numerous facets of
biotechnology are increasingly integrated into the entire
discovery and development process of the "conventional" drug
companies.

Nevertheless, there are a number of new discovery
strategies that have only become possible with the advent of
biotechnology and related tools, and there are a growing
number of biotechnology-derived products being developed as
therapeutic drugs, by either startup biotechnology companies
or by conventional pharmaceutical companies. These are the
endeavors that can be legitimately identified as
biotechnology drug development.

The approach I have taken is to review the information
that has been published about the costs of discovering and
developing conventional drugs; then to decide how we might
expect biotechnology drug development to differ; and finally
to examine whether these ideas are supported by the
information that can be found.

COSTS OF DEVELOPING CONVENTIONAL DRUGS

A new study of costs, by DiMasi, Hansen, Grabowski and Lasagna (1990), has provided up-to-date estimates of the average pre-tax cost of developing a New Chemical Entity (NCE). These authors obtained project-level data on the cost and timing of development from a confidential survey of 12 U.S.-owned pharmaceutical firms for a stratified random sample of 93 self-originated NCEs first tested in humans during the period 1970-82. (The sample was drawn from a comprehensive database of all compounds studied by all pharmaceutical companies in the U.S. since 1962, updated triennially.) R&D expenditure data were collected through 1987 from each firm for each of the phases of clinical drug development, and the preclinical costs for self-originated NCEs were derived by the investigators from a knowledge of each firm's total annual R&D expenditures.

For every NCE that is approved, several others are abandoned at some point. Therefore, the cost of failed projects was allowed for, along with successful ones.

R&D was treated as an investment with returns delayed until marketing approval. To account for the time cost of R&D, the authors capitalized expenditures to the point of NDA approval at a discount rate relevant to the pharmaceutical industry (9% being the preferred value).

Results:

The authors first calculated out-of-pocket costs, namely the total internal plus external costs incurred before adding the cost of capital.

The average out-of-pocket clinical period cost per NCE tested in humans was found to be $11 million (all figures in 1987 dollars). Using an estimated clinical period success rate of 23% (the average value found in the survey) gave an average out-of-pocket clinical period cost per approved NCE of $48 million. Adding estimated preclinical costs of $66M raised the average out-of-pocket cost per approved NCE to $114M.

Development time was estimated for the various phases of development. On average, the time from synthesis to approval was approximately 12 years. Accounting for time costs by capitalizing R&D expenditures at 9% interest rate doubled the total cost per approved NCE to $231M.

Comparing the results to those of a previous study with similar methodology (Hansen, 1979), total development time had increased by two years, and total cost per approved new drug had increased 2.3 times after adjusting for price inflation.

BIOTECHNOLOGY DRUG DEVELOPMENT: SIMILARITIES AND DIFFERENCES

For biotechnology drugs, one could start with the assumption that (very broadly) discovery, preclinical and clinical development might cost about the same as for conventional drugs, while production and quality assurance would cost more.

Let us examine these factors in more detail, identifying in particular the stages where the costs are likely to differ from those of conventional drugs.

From Discovery Through Preclinical Development to
Proof of Concept

Production of Preclinical and Clinical Supplies

Production of active ingredient and formulation into a
pharmaceutical dosage form is an area where biotechnology
drug development costs more than conventional drug
development, since these supplies must be produced in a
manner that meets all applicable standards for toxicology
testing and for the IND (or its equivalent), and in addition
must satisfy all the manufacturing conditions for a sterile
substance. Furthermore, this must be done before the
learning curve and the economies of manufacturing scale have
allowed reduction of costs. Therefore, preclinical and
clinical supplies may be considerably more expensive than
for conventional drugs (e.g., several million dollars up
through the first set of clinical studies).

One consequence of such high cost at an early stage of
development is that it will tend to reduce the number of
analogs in a series that can be brought into clinical
testing.

Delivery Systems

Biotechnology products pose special challenges in
delivery for therapeutic purposes. Unless the native
compound has acceptable pharmacokinetics or is active by one
of the usual parenteral routes, then special dose forms may
have to be developed -- in particular to prolong actions,
protect from degradation, or otherwise improve the
pharmacokinetics.

"Reduplication" of the discovery phase under certain discovery strategies.

Today's biotechnology lead compounds are generally proteins or peptides such as receptors and their ligands, enzymes, modifications of all these, and monoclonal antibodies. (They may in the future be oligosaccharides, as we come to understand the biological role of sugars, or oligonucleotides, as we develop methods of antisense gene modulation. Here I shall only consider protein-related products.)

Biotechnology-derived proteins can lead to drugs in two ways: by yielding medically useful proteins or peptides directly, or by providing tools for discovering the smaller molecules that we know as conventional drugs.

When biotechnology is used to produce protein or polypeptide drugs, we could expect the cost to be in the same general range as for conventional drugs, since the potential savings in a possibly shorter clinical period and higher success rate will tend to be offset by the higher costs incurred much earlier in the process for producing the ingredient and manufacturing suitable pharmaceutical dosage form.

However, pharmaceutical companies prefer small-molecule drugs that are absorbed by mouth and able to be given for long periods for chronic diseases. The costs may be different when biotechnology is used as a tool to discover such molecules. Depending on which strategy is used to exploit the protein-intermediate tools, the costs could be lower or higher than with conventional drugs.

If biotechnology-derived enzymes or receptors are used in vitro to predict (e.g., through the powerful new x-ray crystallography and computational techniques) the structure of synthetic ligands and antagonists and then to screen such compounds for activity, then the cost-effectiveness of an overall discovery facility (which could serve several different therapeutic areas from a single physical installation) might be greatly improved over the older screening techniques.

But another equally valid discovery strategy requires the clinical evaluation of the protein product. Once we look beyond the "replacement" era of biotechnology compounds (e.g., insulin, growth hormone, EPO, and a few others), we have access, through biotechnology, to a large number of potentially interesting natural proteins, plus antagonists in the form of monoclonal antibodies to them, whose therapeutic potentials are unknown and therefore require testing in animal models and eventually man. Before (or while) one embarks on the effort to create the appropriate small-molecule mimic or antagonist of the target protein, it is advisable to determine whether the protein itself is therapeutically useful. So we have to take the large molecule into animal pharmacology and toxicology testing (which is required to be performed but may or may not yield relevant information) and then human trials up through phase 2a, in order to ascertain its therapeutic properties.

Thus for a pharmaceutical company to make full use of biotechnology in a strategic sense, it may have to start by discovering and producing a protein with anticipated activity, then develop that protein through animal pharmacology and toxicology right through the stage of phase 2 clinical trials in order to prove the possibility of therapeutic utility in man. The development process is then started again by treating this protein as a discovery lead for the development of a conventional drug molecule.

Compared with conventional drug development, this strategy involves an additional discovery step which, requiring substantial investment so early in the drug development process, will increase the overall costs of drug development, perhaps substantially.

On the other hand there are in this strategy factors that would offset this higher cost. They include:

o   Bringing the intermediate protein itself to the market and generating enough profit to recover the costs of the additional discovery step.

o   Generating more discoveries, potential leads, and development compounds from the same discovery program.

o   Increasing the final success rate of the development compounds that result from this discovery strategy.

It is too early to tell whether these cost offsets will indeed occur.

Clinical Development and Regulatory Approval

Assuming that the standards of regulatory approval are unchanged for biotechnology drugs, it is possible that because of their parenteral nature (which leads the compound into acute or short-term treatment indications), the length of the clinical period from the start of phase 1 testing to submission of the PLA (Product License Application) at the end of phase 3 could be less, and hence the costs for the clinical program (excluding supplies) could be less than for

conventional drugs. However there is not enough experience
to date with biotechnology PLA submissions and approvals to
determine whether this is so.

Manufacturing

Manufacturing costs are a major issue in biotechnology
and the process has been described as the industry's
Achilles heel (Boss, 1990). By contrast with conventional
drugs in an established pharmaceutical company, choice of a
manufacturing strategy--particularly for a biotechnology
startup company--is relatively complex and could be critical
to the success or survival of the company.

For a conventional drug, an amortized plant is likely to
exist or capacity can be hired, since there is a worldwide
surplus. By contrast in the case of a biotechnology product,
the facilities do not exist and probably have to be built
from scratch at a cost of, say, $50M.

In the case of a fairly simple biological, such as a
monoclonal antibody, Boss (1990) estimated that the fixed
cost would be $20 million, while the manufacturing cost per
gram would be $2,200 before depreciating the capital costs.
The price of the final product to the patient was estimated
to be $4,300 gram assuming a 30% cost of goods. It was
further noted that a course of therapy for an anti-cancer
monoclonal could require several doses of up to 10 grams
each, with a marginal manufacturing cost of active
ingredient of up to $60,000. These costs and prices are
much greater than the price of therapy with conventional
drugs. It should also be noted that some experts familiar
with the area have estimated that the above costs per gram
are a substantial underestimate, or could be achieved only
after a long learning curve of production.

Costs of Marketing

There are few estimates of the costs of marketing biotechnology products specifically. However, marketing costs are not affected by the origin of the product, but depend on the therapeutic area, the exact properties and indications the new drug has, and the nature of the company (e.g. established or startup).

We shall therefore take a typical biotechnology product, namely a recombinant protein and make the following key product assumptions:

- o Hospital use
- o IM/IV administration
- o Relatively narrow physician audience (e.g., infectious disease or surgeons)
- o No samples are distributed
- o Enters an already established therapeutic category, eliminating the need for broadly-based educational efforts.

For a product with these characteristics, the initial marketing costs were estimated to be $7 to 8 million per year, with a total of $23 million for the first three years.

Other Factors

Protection of Intellectual Property

While the costs of patenting biotechnology products may be a relatively modest direct cost of drug development, a more important question is the amount of protection thus obtained, because this is a powerful determinant of whether biotechnology investments can be protected, and hence of the attractiveness of this investment area.

There are no patent laws specifically for biotechnology products, and new issues and interpretations have arisen with the types of discoveries made in biotechnology. In conventional drug development by the established pharmaceutical companies, there is now enough history and knowledge of the issues for disputes to be settled without litigation, usually by cross-licensing agreements. By contrast in the biotechnology area, the merits are less well defined and disputes tend to involve litigation. This is because of the very high stakes involved for the individual companies (particularly in the case of the biotechnology startup companies), their extreme dependence on one or two initial products, and the independent personalities of the leaders attracted to startup companies in this frontier of science and business (Mertz, 1990).

Examples of issues in biotechnology that have gone to litigation include the action by Genentech against Burroughs Welcome and Genetics Institute over three patents for TPA; Xoma's suit against Centocor over antibodies to endotoxins, and between Amgen and Genetics Institute over recombinant erythropoietin.

Until the results of the various disputes are settled by litigation or agreements and we have a general knowledge of the rules that result, there will continue to be less certainty about the extent of intellectual property protection in the biotechnology area than in conventional drug development.

EXAMPLES OF OVERALL COSTS IN BIOTECHNOLOGY

Interleukin-2

One detailed example is available in the literature, and this suggests that the costs of developing a biotechnology protein drug are indeed fairly comparable to those of developing conventional drugs. For the development of Interleukin-2, M. Ostrach of the Cetus Corporation stated in December 1989 that Cetus' RD&C costs had been $75M before including the cost of capital. In addition, the pilot plus full-scale manufacturing facilities cost $45M, and marketing $15M.

The $75M figure is the one that should be compared with the DiMasi et al estimate of $114M, before interest, for conventional drugs. As IL-2 has not yet been approved for the U.S. or other major markets, and since Ostrach estimated that a one-year delay in introduction would add $35M to the development cost, the total would reach $110M by the end of 1990. Thus, the additional development time and costs needed to reach the major markets, plus the cost of capital, will bring the total time and cost of developing IL-2 well into the range of the average conventional NCE.

An alternate calculation was to take Cetus' total costs over the 10-year development period, to that date, of IL-2. This method shares the advantage of DiMasi et al of including "dry wells", but since the outcomes are not yet known it is probably an overestimate if attributed solely to IL-2. By this method, $370M had been incurred (by December 1989, before any product had been approved for the U.S. market.)

DISCUSSION AND CONCLUSIONS

The Problem:

While biotechnology has transformed the discovery phase, it will not necessarily reduce the cost of drug development. Based on the very limited information available to date, it appears that the cost of developing individual biotechnology protein drugs will be fairly similar to that of conventional drugs, namely over $200M per NCE on average, including the cost of capital which is half the total. There are several factors that could move this estimate up or down, and a more precise value will have to await a larger sample of approved biotechnology drugs and a special study of the type done by DiMasi et al.

Although the regulatory environment for biotechnology drug development began with less encumbrances than today's conventional drug development, it is now tending to show some of the characteristics of the mature regulatory environment that surrounds conventional drug development, and this tendency may increase in the future. Furthermore, the progressive increase of regulatory standards and requirements means that costs are likely to rise also.

It should be noted that because we are still in the "startup" phase of biotechnology drug development, the true average costs of biotechnology drugs will not be assessable until a more steady state is reached, in particular until the fate (success or failure) of the first generation of development candidates has become known. This will not occur before the mid 1990s at the earliest. Until then, the development time, costs and success rates of those biotechnology drugs actually approved will tend to be more optimistic than the real average, although by steadily diminishing amounts. This is due to several factors: the

first generation of biotechnology products consisted of the
most obvious and logical targets, (e.g., replacements for
hormones or other substances already available from natural
sources, such as insulin, growth hormone and alpha
interferon); the examples approved first necessarily include
those with shorter development times; and the regulatory
requirements for this new technology were least at its
inception. It is only after the mid 1990s, when this
startup phase is complete, that we will be able to measure
the true mean times and costs.

The more precise mechanistic targeting of biotechnology
products should logically lead to drugs that are more
precise than their predecessors, with more specific
efficacy, less toxicity and a sharper focus of drug effects.
At the same time, however, this very specificity could
narrow the patient population for whom the drug is approved
and indicated and hence the size of the market--perhaps even
to "orphan" indications--while the costs of these products
could raise questions of reimbursement, including what
indications are reimbursable under the various cost-
constraint strategies that are being developed and
tightened.

While this may be an inevitable cost of the increasing
depth of the discoveries we can now make with biotechnology,
it does mean that all who are concerned with the drug
discovery process (not only the companies involved, but also
academic researchers and regulators) will need to keep at
least an open mind to prevent the utilization of these new
discoveries from becoming even more costly and time-
consuming. Today's constraints could lead to an impasse in
the development of drugs from biotechnology, or at least
retard the transfer of the discoveries of biotechnology into
available therapies.

Possible Solutions:

The necessarily high relative cost of biotechnology products, and the small target patient populations for some of them, suggest that in this era of cost constraints, some new thinking is needed to optimize the conditions for commercialization of new potential therapeutic discoveries from the biotechnology sector.

From the biological and industrial perspectives, much remains to be learned about the most efficient discovery and development strategies. (We may even have pleasant surprises as the wealth of biotechnology options makes discovery easier; but realistically, on past trends, it is more likely that costs will increase.) To be successful a pharmaceutical company must ascertain the most cost-effective strategies and ensure they are followed, but it will be some time before the optimal courses become clear.

The regulatory environment is a major area that is under society's control, since it is the height of the regulatory hurdles to clinical investigation and marketing approval, plus the length of the regulatory review time, that play a large role in determining both the time and the costs of drug development.

There have been numerous studies of how to optimize the regulation of pharmaceutical products (Hutt 1984), and another high-level study on the Food and Drug Administration, has been set up in the U.S. (the Health & Human Services Advisory Committe on FDA) before the previous one (the National Committee to Review Current Procedures for Approval of New Drugs for Cancer and AIDS) had even completed its report. By now the measures that could be taken are well known (PMA, 1990). Some of the key points I

believe are most relevant to biotechnology products can be
summarized as follows:

- o   Rationalize regulatory requirements medically.
      There is little point in perpetuating regulatory
      requirements that are not strictly necessary
      medically. A large simplification could be achieved
      if all regulatory requirements were reassessed by
      this criterion to prevent the accretion and
      perpetuation of unnecessary requirements. This is
      one of the few areas where progress is possible, and
      one with a large effect. Di Masi et al showed that
      a year's reduction in the duration of phase 3 would
      reduce the capitalized development cost by $18M,
      while Ostrach showed that a year's delay with IL-2
      would cost $35M.

      An example of rationalization would be simplifying
      the regulatory requirements for certain classes of
      products, such as mouse monoclonal antibodies.
      There is now considerable experience with these
      compounds in man, and they appear to be relatively
      non-toxic. Can ways be found of simplifying the
      quality assurance standards, and even deleting the
      toxicology requirements needed, before certain types
      of human trials are permitted?

- o   Harmonize rational requirements: EC-US-Japan.
      There are three major world regions for drug
      development and marketing: the EC, the USA and
      Japan. This is a critical moment in history as the
      EC seeks to harmonize the regulatory system for its
      internal pharmaceutical market. To avoid further
      increasing the costs, it is essential that such
      regulatory requirements, while being rationalized
      are also harmonized internationally. If these two

steps do not occur, precious drug development
resources will continue to be wasted and drug
development costs will be unnecessarily high.

o   Good Regulatory Practices. A considerable
improvement in efficiency might be achieved if
regulatory agencies were required to adhere to
generally recognized management standards, such as
meeting deadlines, answering correspondence
promptly, respecting agreements, and holding
managers accountable for the performance of their
groups.

o   Substituting phase 4 studies for part of phase 3.
In view of the real difficulties in finding ways to
shorten or truncate the increasingly-burdened system
of drug development and approval, it has now become
more respectable to think of an idea that was
previously considered too radical: truncation or
elimination of phase 3 studies in favor of phase 4.
This, along with shortening the approval process is
the only place where really substantial time, and
hence resources, can be saved (e.g., a total of five
years). The arguments are contained in the above
references and are too long to consider in detail
here. However, devising an acceptable method of
achieving this may be one of the constructive and
rewarding challenges of the 1990s.

294

REFERENCES:

Boss M. (1990) Biotech's Achilles Heel: Manufacturing is Risky and Expensive. In Vivo 8:1-4

DiMasi J, Grabowski H, Lasagna L, Hansen R (1990) The Cost of Innovation in the Pharmaceutical Industry. (Submitted)

Hansen R. (1979) The Pharmacological Development Process: Estimate of Development Cost and Times and the Effects of Proposed Regulatory Changes. Chien RI (Ed) Issues in Pharmaceutical Economics. DC Heath & Co., Lexington p.151-187

Mertz, B. (1990) Suit Outcomes Will Set Precedents: Biotech Firms Battle Over Patent Rights. American Medical News May 4:1 (Newsletter)

Ostrach, M (1989) The Cost of Developing Medicines. Insight for Alternative Futures Foresight Seminar, Alexandria, VA

Pharmaceutical Manufacturers Association Statement on the Food and Drug Administration to the Health & Human Services Advisory Committee Sept. 1990 PMA, Washington D.C.

Hutt P B, (1984) Investigations and Reports on the Food and Drug Administration. 75th Anniversary Vol. of Food and Drug Law 27 (1984), Food and Drug Law Institution

Discussion - COSTS OF DISCOVERING, DEVELOPING, MANUFACTURING AND
MARKETING BIOTECHNOLOGY PRODUCTS

L. Gauci

One should not forget that the very rapid development of interferon alpha for the treatment of hairy cell leukemia indicates the willingness of regulatory agencies involved, to assist in the registration throughout the world.

W.M. Wardell

But that was one of the first compounds of the new era, and new questions will be asked of subsequent ones that were not asked of the first. I think it is inevitable that there will be an accretion of requirements. That is what has happened with conventional drugs, and my impression is that it is happening with biologics as well.

P. Juul

I am not very happy with a registration after phase II, as it has been suggested in Europe concerning certain anti-cancer drugs, and I am not in favor of letting the expanded access program, which we also have problems with, lead to an earlier registration. I think there is another solution. If one accepts compassionate IND usage then it is acceptable that the drug is only used by specialists but we can accept that hospitals or the patients pay for the drug, meaning that you will have an interim phase when the product is not registered but the company is not loosing these enormous sums of money which you mentioned. One of my reasons for not accepting or liking an early registration is that it is our experience that it is very difficult to get rid of a drug once it is on the market: The regulatory authorities have as much difficulty in getting rid of it as the company originally had in getting it into the market. So these patients even with serious diseases should not be treated more badly than other patients. With regard to good regulatory practice I would say that U.S. and Japan could just follow the rules of the EEC where NDA should be definitely answered within less than a year and if we don't do it and we don't answer, it means that we have to accept the drug on the market.

M.M. Reidenberg

It has been shown that virtually all of the drugs that get well into phase III end up getting marketed. There is also data showing that phase III has too few patients in it to pick up adverse events that occur as frequently as one in a thousand. So, I think that a really thoughtful consideration of what information is learned during phase III that is essential for early marketing would probably show that most of the time very little is learned. I think that as a society we make a fundamental mistake in thinking that a drug is experimental on day minus one; on day zero it is approved as safe and effective, and on day plus one it become standard practice and it is a safe and effective drug. Our perceptions as a society, our de facto regulatory requirements, are totally out of congruence with the reality of therapeutics. I feel that it would be far preferable to have drugs made generally available much earlier in the development process with the understanding that they really are experimental. I think with this we would need to have an agreement for more substantive scientific research after marketing. And I think it would be necessary for the regulatory agencies to be able to reasonably reevaluate decisions for marketing and labelling, and have the ability to go all the way to removal of an approved drug. I think that the issue then for the company is how much risk to take that rescinding approval may occur. I suspect most company's managements would oppose it, but I think such a process would bring regulatory and developmental activities far closer to the reality of therapeutics.

L. Gauci

If there was a mechanism whereby the drug could be sold to patients suffering from conditions for which it is not properly developed, this would allow the company to finish the work properly and may help on improving dosaging schedules.

D. Maruhn

I would like to take issue with the notion that on an average clinical development costs of biotechnology products might be lower than those of conventional drugs. That could be true for drugs which are used for short term treatments but in the case of substitution therapy the costs of clinical development are similar to those of conventional drugs. However, one should take into account that reduplication of clinical studies certainly adds a substantial burden to the costs of development. We are trying to optimize our international efforts and what we use is an instrument what we call the International Clinical Development Plan thus trying to avoid that too large number of patients are exposed to the drug and to ensure that we get the minimum of requirements for all important countries where we are going to apply for a registration of the drug.

© 1991 Elsevier Science Publishers B.V. (Biomedical Division)
The clinical pharmacology of biotechnology products.
M.M. Reidenberg, editor

THE IMPACT OF ECONOMIC ISSUES ON THE THERAPEUTIC USAGE OF
BIOTECHNOLOGY PRODUCTS. A VIEW FROM THE HOSPITAL

MICHA LEVY AND SAMUEL PENCHAS
Hadassah University Hospital, P.O. Box 12000, Jerusalem
91120, Israel

BACKGROUND

Biotechnology products at the moment comprise about 1% of
the $130x10^9$ worldwide drug market (1). According to a
recent estimate U.S. sales of biotechnology-derived products
are expected to triple from $900 million in 1989 to $3000
million in 1993 (2). Current FDA-approved products which
include tissue plasminogen activator (t-PA), Alpha-
interferon, human recombinant insulin, growth hormone and
erythropoietin and hepatitis B vaccine are expected to
achieve a $1950 million U.S. market size by 1993.

TABLE I

FDA-APPROVED BIOTECHNOLOGY PRODUCTS
Expected U.S. market size in 1993

	$ million
t-PA	400
Alpha-Interferon	350
Human Insulin	250
HGH	200
Erythropoietin	600
Hepatitis B Vaccine	150
Total	1,950

(Adapted from ref. 2).

About twenty other products will become available by the mid-
nineties. These will include activated protein C and factor
VIII-C, Superoxide dismutase, beta and gamma interferon,
interleukins, growth and wound-healing factors, colony

stimulating factors, malaria, herpes, hepatitis C and, hopefully, AIDS vaccines and a variety of monoclonal antibodies. More than a hundred products are now undergoing clinical trials. In addition, a colossal market is expected for new biotechnology-derived biosensors, equipment and instrumentation (3).

Turning from these grandiose images and their stock-market counterparts to our present hospital pharmacy budget reveals a surprisingly modest impact of biotechology products. Is it the tip of an iceberg or just flakes before the avalanche? The answer is both.

The Hadassah University Hospital in Jerusalem, Israel, has 876 beds. It serves as both a community hospital for the half-million residents of the city, and as a tertiary-care center and oncological center for the country. It performs kidney, heart and bone-marrow transplantations. Its' annual budget amounts to $110 million and the pharmacy budget to $9 million, out of which $8 million pays for purchase of drugs.

Current Use of Biotechnolgy Products

All FDA-approved biotechnology drugs are registered in Israel and all but t-PA and Alpha Interferon approved for hospital use. However, thus far the only biotechnology product for inpatient use included in the Hadassah Hospital formulary is human insulin, which comprised 2/3 of the $21,000 hospital insulin purchase in 1989. Altogether the price of insulin has remained stable.

Fibrinolytic therapy mainly with streptokinase and urokinase, was administered last year to about 250 patients at a cost of $75,000. Our cardiologists and the P and T Committee follow with great concern the ongoing clinical trials comparing t-PA with streptokinase. Though several such trials have reported more rapid and frequent coronary artery thrombolysis following t-PA (4,5), more recently the results of the GISSI II Study (6) concerning in-hospital mortality (9.2% in those treated with t-PA + heparin vs. 7.9% for SK + heparin) do not warrent a non-reimbursable expenditure to the hospital of $0.75 million per year, to be caused by the introduction of t-PA. Hadassah does not at this time purchase t-pa and routinely supplies only streptokinase and urokinase (for later follow-through).

The Hospital has approved screening for HBc antibodies and vaccination of its' 4,600 personnel with <u>recombinant hepatitis B virus vaccine.</u> The cost of screening ($10 per person) and vaccination ($16.5 per person) amounts to $107,380. The Hadassah administration has attempted to fund these expenses from various non-budgetary sources (donations, grants). It is predicted by our hepatologist that this program will eventually prevent the development of about 80 cases of cirrhosis and 8 cases of hepatocellular carcinoma (7). Recombinant erythropoietin, growth hormone and alpha-interferon are provided on ambulatory basis directly supplied by the insurers' pharmacy and funds.

In Jerusalem there are about 170 patients with end-stage renal disease, 35 are treated with <u>erythropoietin</u> (rHuEpo) at an average cost of $5,000 per patient per year. The Ministry of Health, has issued vigilant prescribing criteria in line with the guidelines of the American National Kidney Foundation (8). The benefit to the patient appears to be very favourable, including transplantion of patients who were previously considered to be untransplantable. This cost is covered by each patient's insurer (sick-fund).

<u>Human growth hormone</u> is prescribed following approval by experts of the Ministry of Health, to 10 Jerusalem children at an annual cost of about $10,000 each. Natural human growth hormone is no longer used. Hairy-cell leukemia is thus far the only Ministry-approved indication for <u>Alpha-interferon.</u> (1-2 patients annually). About a dozen patients receive the drug on experimental protocols for chronic myeloid leukemia, lymphomas and viral hepatitis. The cost, covered by research grants or by the patients themselves, was estimated at close to $70,000 last year.

Beta-interferon, IL-2 and CSFs are being used in the hospital as part of experimental protocols approved by the IRB and the Ministry of Health. For the time being the drugs are provided by the manufacturers, free of charge. OKT 3 was used on two patients who received the drug from abroad (cost $10,000). In addition, monoclonal antibodies and molecular probes are used for diagnostics. The annual expenditure

amounts to $280,000 Thus, although the direct hospital
expenditure on biotechnology drugs is only a modest $15,000
(or less than 0.2% of the pharmacy budget), the real cost at
present of biotechnology products for our in-patients and
out-patients treated in our hospital is close to 700,000
dollars annually (8% of the pharmacy budget).

TABLE 2

EXPENDITURE ON BIOTECHNOLOGY PRODUCTS
Hadassah's Patients and Personnel (1990)

	$
Human Interferon	15,000
Alpha Interferon	70,000
HGH	100,000
Erythropoietin	140,000
Hepatitis B Vaccine	107,000
t-PA	none
Experimental products	gratis
Antibodies & probes	280,000

Another consideration is of course the money-saving
properties of biotechnology products. Cure, prolongation and
better quality of life, productivity and decreased need for
medical services are all of paramount importance from the
general medical point of view. For the hospital budgetary
system, financial gains appear to be modest and relate to
the operating reimbursment systems. To give one example,
according to our DRG system the hospital is being paid
$15,000 for a bone-marrow transplant. It is expected that
the use of CSFs will shorten the length of hospitalization
and decrease the cost to the hospital by 20%.

Future use of biotechnology products

The present situation does not forecast the future economic
impact of biotechnology products. Many of the drugs
currently undergoing clinical trials will, unlike the
competitive price of human insulin or the rareness of
indication for HGH, enjoy the expensive combination of being

both unique and of relatively wide applicability. Moreover,
biotechnology products are expected to reach the market much
faster than conventional drugs (i.e. < 15 years) (3).
Herfindal (9) has recently described the "beachhead" strategy
of the industry, namely registration of the product for a
narrow indication, followed by unlabeled drug use, allerting
the general and medical public and creating pressure on the P
and T committees. (A good example is the use of interferon).
Unlabled use is of particular concern to hospitals as third-
party payers tend not to reimburse such use.

Strategies for dealing with biotechnological innovation

Herfindal (9) suggested the following strategies:

* Implement prescribing protocols and guidelines
* Analyze impact on other treatments
* Analyze economic impact
* Implement monitoring and surveillance programs

As for analysis of economic impacts, his experience at the
University of California, San Francisco, is being cited.
Requests for high-cost drugs are handled by the hospital
administration as major capital purchases, similar to the
opening of new clinical services, or the implementation of a
heart transplantation program. Complete economic, clinical
and strategic justification are required in such a process,
including cost, reimbursement, impact on other services and
personnel requirement. He concluded that for a pharmacy
department it is impossible to absorb the cost of high-cost
drugs without shift of resources and that the ability to pay
for the new therapeutic modalities will quickly be outpaced
by the industry's ability to produce them. We agree with
these forecasts and would like to expand on the application
of hospital strategies.

The informed public is demanding a rational and consistent
program for the appropriate use of the limited health
resources. Priorities must be selected in medical care and a
scale of "reasonable requirements" established (10). In
seeking a scale of "reasonable requirements" in the hospital
setting, health-care providers, clinically and
administratively involved with hospitalized patients must
formulate a flow-chart of averaged, relative, weighted,

multiple parameters bearing on the main aspects of the patients' health to replace those of arbitrary triage by financial resources, age of patient and delay. All parameters should comprise measurable elements of technology: equipment, devices, drugs, procedures and capital and human investments. Each parameter along the scale must be regularly assessed for its efficacy, safety and relevance in the provision of hospital care - both in specific situations and in the wider social and ethical spheres. Nine parameters compose the suggested scale. They are: 1) the net therapeutic contribution of the procedure - the degree to which it will preserve life, limb or function; 2) the reduction of physical and/or mental suffering; 3) improvement of the quality of public health - both in prevention of disease or injury, and in promoting the effectiveness of health services and community health practices; 4) the balancing of therapeutic against diagnostic procedure; 5) diagnostics - perfecting its positive predictive value; 6) therapeutics - achieving objectives, which include: elimination of pain, easing and speeding maximal recovery and rehabilitation; 7) costs - direct, indirect and capitalization; 8) development program, its potential contribution in the future; 9) hazards of medications both nonspecific and unidentified.

These complex parameters, together with further potential facets, are naturally subject to periodic variations, to overlap and merging, and to changing trends of science, morbidity, utilization and economics. They therefore have to be correctly weighted in the particular algorithm relevant to each case.

They are proposed as the basis on which to build a systematic methodology for the provision of hospital biotechnology innovations. The target of any such methodology should be an objective weighting algorithm for regulating the decision-making process - one that is capable of accomodating both anticipated and unforeseen problems.

Parameters for the evaluation of the economic impact of biotechnology products in the hospital setting.

The net therapeutic contribution of the product is of

course of paramount importance; so is the reduction of suffering in any medical setting. The third parameter is the improvement in the quality of public health which relates perhaps most to the post-hospitalization state and should not be underestimated. Hospital staff should not take the narrow view of the patient in a time-limited stay. The fourth parameter will exhibit itself forcefully with the increased introduction of biotechnology products. The fifth point relates particularly to the use of monoclonal antibodies and molecular probes for diagnosis, while the sixth point is obvious in any case. The seventh parameter, cost, will be discussed separately.

As to the eighth parameter, hospital biotechnology drug development schemes should be evaluated on the basis of their anticipated duration and estimates as to their final contribution. Lastly, the hazards of drugs - particularly by biotechnology methods, have to be taken into account

The Assessment of Cost

When facing costs, hospital administrators tend to look at an issue differently if its costs are met through the regular operating budget as against any other "non-budgetary" source (i.e. donations, grants, contracts, patents etc.). If the utilization of particular new biotechnology products are considered budgetary - the hospital administration will institute a terrific battle to recover the outlay from regular budgetary sources (i.e. insurers, third party carriers, governmental subsidy or the patients themselves). To this end, the administration will use various pressure techniques to force the payers to pay. This could be by insinuating to the patients that the specific product was absolutely essential or at least far better for their particular need, and then use a phalanx of fearful and agitated patients and their families to force an agreement from a payer. The beachhead technique described by Herfindal (9) is well known to hospital administrations; get the payer to agree that interferon is the drug of choice and to be paid for hairy cell leukemia and thus introduce it to the hospital formulary and then use the salami technique extending its use slice by slice - indication by indication.

Obviously in parallel with the battle which the hospital administration, allied with its' clinicians, wages against the payer - it wages an internal front against its own clinicians. The attempts by the clinicans to introduce new biotechnology products are, except in a few extraordinary uses which can boast of exemplary scores on each of the nine parameters, resisted by the administration ferociously. Before agreeing to a budgetary acceptance management will try to insist that the new drug be paid for from the clinicans' research funds (being "his own baby" and not an accepted drug) or from a grant if the clinican is in anyway at all connected to a manufacturer or other large research institute. In most Western countries the official certification of drugs, initiated of course primarily for safety reasons, hands management an oportunity for delaying and procrastinating for the introduction of biotechnology products: the process is long and even after formally completed a lot of small details remain to be completed. These multiple steps enable a determined management to resist an increase in its drug purchasing line in the budget, mainly by delaying through officiousness.

CONCLUSIONS

The introduction of biotechnology products involves the same boons and banes of all new high tech. The blessings are extremely powerful effects, never thought of even a few years ago. The curse lies in the cost: similarly to other high tech developments, these products are, initially, with rare exceptions, each individually expensive. In the climate where cost containment is the overlying grand motto of todays' medicine and when a letter to the New England Journal of Medicine suggests that the contemporary physician's Oath will contain "I will always use generic drugs" (11), expensive medications suffer the same travail of being approved, accepted and made available as other expensive medical products. A certificate-of-need legislation pertaining to drugs has not yet been suggested, to the best of our knowledge, but administrative reluctance leading to administrative obstruction is widespread and perhaps natural.

How does one overcome all this? We suggested the

application of the scale of "reasonable requirements" using
the check list of nine parameters in an atmosphere cognizant
of there being limits to available resources and that it is
the duty of clinicians, researchers, administrators and
manufacturers to priorize with moral responsibility.

ACKNOWLEDGMENTS
Drs. Gotsman, Jacobs, Livshits, Rubinger, Shouval and Slavin
provided helpful data and advice.

REFERENCES

1. Roche to buy leading U.S. biotechnology firm.   The
   Pharmaceutical Journal 10.2.1990 p. 177

2. U.S. Biotechnology products - $3000 million by 1993. SCRIP
   8.6.1990

3. Werner RG Biobusiness in pharmaceutical industry. Arzneim.
   Forsch/Drug Res 37:1086-93, 1987

4. TIMI study group. The thrombolysis in myocardial
    infarction (TIMI) trial. N Engl J Med 312:932-36, 1985

5. Collen D Coronary thrombosis: Streptokinase or recombinant
   tissue-type plasminogen activator. Ann Intern Med 112:529-
   38, 1990

6. The International Study Group In-hospital mortality and
   clinical course of 20,891 patients with suspected acute
   myocardial infarction randomized between alteplaze and
   streptokinase with and without heparin. Lancet 336:71-5,
   1990.

7. Shouval D Personal communication.

8. Ad Hoc Committee for the National Kidney Foundation
   Statement on the clinical use of recombinant
   erythropoietin in anemia of end-stage renal disease.   Am
   J Kid Disease 14:163-69, 1989

9. Herfindal ET Formulary management of biotechnological
   drugs. Am J Hosp Pharm 46:2516-25, 1989

10.Penchas S Toward the development of a technology
   assessment methodology in medical diagnosis and therapy.
   Isr J Med Sci 22:201-2, 1986

11.Manuel BM A contemporary physician's oath. N Engl J Med
   318:521-2, 1988.

Discussion -THE IMPACT OF ECONOMIC ISSUES ON THE THERAPEUTIC USAGE
OF BIOTECHNOLOGY PRODUCTS. A VIEW FROM THE HOSPITAL

M.M. Reidenberg

Do you think that the improvement of quality of public health is an issue that should be dealt with by a formulary committee on whether to include a product, or do you feel that an individual physician treating an individual patient should weigh the value for the particular patient against what it does to society as a whole?

M. Levy

I think that for the hospital committee it is a general issue. I should not refer to the individual patient.

B.R. Meyer

As a chairman of pharmacy and therapeutics committee I think you have summarized very well the problems that we face. The numbers in our hospital are very similar to yours. We however have spent considerably more on erythropoietin where we spend about $400,000-500,000 a year. We have an extensive renal dialysis programme and we get about 75-80% of that back by third party payers. We also spent an awful lot of money on something you didn't mention which is i.v. gamma globulins

L. Gauci

I would like to refer to your list of reasonable requirements. I cannot agree with the notion that biotechnology products should be considered in a separate category because they are the most expensive drugs. This is not true, there are other drugs which are very expensive. I think that in the case of modern drugs, which are expensive to develop, pricing considerations should be included in the phase III studies.

M. Levy

I agree. As a group the most expensive drugs known to hospitals are antibiotics. We spend between to $2-3 million a year on sophisticated last-generation antibiotics.

© 1991 Elsevier Science Publishers B.V. (Biomedical Division)
The clinical pharmacology of biotechnology products.
M.M. Reidenberg, editor

BIOTECHNOLOGICAL PRODUCTS, CAN THIS COMMODITY BE AFFORDED?

FERNANDO GARCIA ALONSO
Ministry of Health and Consumers Affairs, Paseo del Prado 18-20,
28014 Madrid (Spain)

INTRODUCTION

There is an important gap between basic biotechnological research and the subsequent product application in clinical practice. Thanks to this basic research it is possible to obtain products such as Growth Hormone, Interleukin, Interferon, Somatostatin or Erythropoietin. As soon as they are obtained a long searching process for clinical indications starts. In this process a conflict between four agents is generated:

- The developing Pharmaceutical Company, that would like to quickly extend the authorized clinical indications .
- The potencial prescribing physicians, who would like to use the new product in indications that are still waiting for the clinical trials' results.
- The Regulatory Authority, that tries to be quite sure before approving new indications.
- The National Health Services (NHS), that don't have funds enough to finance these new and very expensive products.

This clash of interests arise from curious situations that would be interesting to analize.

THE CASE OF GROWTH HORMONE

Short stature was treated with Growth Hormone (GH) obtained from human pituitary gland in 1958 for the first time (1). In 1985 the original biosynthetical GH was introduced after several young men who were treated in their childhood with GH died of Creutzfeldt-Jakob illness (2). This new GH has allowed not only a more extensive use, but also the exploration of potentially new indications. The current approved indications are: defficiency of GH secretion, neurosecretory dysfunction, inactive GH and Turner syndrome (3).

Other pathologies that could be new indications for GH treatment in the future are: constitutional short stature, intrauterine growth retardation, renal insufficiency, osteochondrodysplasias and Prader-Willi syndrome. GH could also be used as a simply aesthetic factor in men over 60 years old. This new perspective arises from a recently published study showing that the disminished secretion of GH is partially responsible for the decrease of lean body mass, the expansion of adipose-tissue mass, and the thining of the skin that occurs in old age (4).

This situation envisages a new GH expansion in the market and its limits are impossible to predict. In Spain 65.300 vials of GH were prescribed during 1986, 367.000 vials in 1987, 885.000 vials in 1988 and 1.387.784 vials in 1989 (5). This represents a $138 million expenditure in 1989 for the Spanish NHS in spite of the fact that this drug is "hospitalary diagnostic" (a category that forces the prescription of this drug to be made under a medical report that has to be checked by a Health Inspector). To evaluate this magnitude, it's interesting to know that the total drug expenditure on primary health care financed for the Spanish NHS was $4000 millions during 1989 (6).

Bearing in mind these data, the affordability of this commodity has been possible so far but probably won't be in the very near future. Perhaps it should be taken into consideration if it must be afforded. In a society where resources are limited, these must be used in an efficient way. What is expended in GH is to the detriment of other health resources. Deciding what clinical indications must be approved, or the economical limits on GH financing is very difficult, because physicians and health planners don't usually agree on these matters.

In any case, we have to admit that the current situation is not good and reveals the lack of adjustment between the quick development of biologically obtained products and established drug evaluation systems. It is curious to observe how the GH consumption in Spain presents great disagreements when compared with consumption data from other European countries. Regarding the 1988 data, there were 152 patients per million Spanish inhabitants treated with GH , whereas there were 38 in Germany , 37 in the United Kingdom, 67 in Holland, 100 in France and 88 in Italy (7). Even admitting that achieving this reckoning is difficult and that involves some error, there appears to be differences between European Countries that are hardly justifiable. The same occurs when the data from the seventeen Spanish Autonomous Regions are analized. When adjusted per population, results indicate that some Regions consume GH eight times more than others.

THE CASE OF INTERFERON

Currently there are three Interferons in Europe developed by three different Laboratories. The obtaining mechanism changes from one to another but, at least theoretically, we could not expect high differences concerning their clinical efficacy. In spite of this, the indications that each laboratory applies for are not coincidental. Particularly we will analyze Interferon alfa 2b. Table I shows that the Company has applied for up to 10 different clinical indications along the 12 EEC countries, that the 12 different Regulatory Authorities have accepted in 12 different ways, that is to say that there are not even two countries accepting the same conditions. There seems to be a relative accord on Hairy cell leukemia and AIDS-related Kaposi's Sarcome, but for the indications left, it seems difficult to find a laudable explanation.

Another remark deserves the Hepatitis case, because the frequency of this illness could represent a hardly bearable economical charge in countries where the appropiate indication is approved. In November 1989 the use of Interferon alfa 2b, was approved for Hepatitis B and C in four countries (Greece, Ireland, Italy and Portugal). The eight countries left are still deciding. In this situation two conjectures can be made:

a) The four countries that approved this indication have a slightly less strict regulation system.

b) The four countries that approved this indications have an efficient regulatory system that response quickly to scientific evidence.

Which of the two conjectures is true, if any?. In case of doubt, we leave it to the reader's criterium. It is curious to know that in November 1989, two clinical trials were published in the same issue of The New England Journal of Medicine, and both of them conclude that a 24 week course of Interferon alfa 2b therapy is effective in controlling disease activity in many patients with Hepatitis C, although relapse after the cessation of treatment is common (8) and that Interferon alfa 2b therapy is beneficial in reducing disease activity in chronic hepatitis C, however the beneficial responses are often transient (9).

Regarding Hepatitis B, more recently another clinical trial has been published and the author concludes that treatment with

TABLE I

REGISTRATION STATUS OF INTERFERON ALFA 2b FOR THE DIFFERENT INDICATIONS IN THE EEC (NOVEMBER 1989).

Approval date for following indication	BE	DK	GE	GR	SP	FR	IRL	IT	LUX	NL	PO	UK
Hairy cell leukemia	85	86	87	87	88	88	85	87	86	86	86	86
AIDS-related Kaposi's Sarcoma	85	86	–	87	–	86	85	87	86	–	86	87
Multiple Mieloma	85	–	–	–	–	–	85	87	86	–	86	–
Laryngeal Papillomatosis	85	–	–	–	–	–	–	–	–	–	–	–
Basal cell carcinoma	89	–	–	–	–	–	–	–	–	–	89	–
Non-Hodkin's Lymphoma	89	–	–	–	–	–	–	89	–	–	–	–
Cutaneous T-cell Lymphoma	89	–	–	–	–	–	–	89	–	89	–	–
Chronic mielogenicus leukemia	–	–	–	–	–	88	88	89	–	–	89	87
Bladder Cancer	–	–	–	–	–	–	–	–	–	–	–	–
Condyloma Acuminata	–	86	–	–	–	–	85	89	–	–	86	87

Interferon alfa 2b (5 million units per day for 16 weeks) was effective in inducing a sustained loss of viral replication and achieving remission, assessed biochemically and histologically, in over a third of patients. Moreover, in about 10 per cent of the patients treated with Interferon, hepatitis B surface antigen disappeared from serum (10).

This three studies are a good sample of state of the art Hepatitis treatment with Interferon. Its seems very clear that there is a therapeutic benefit but with important restrictions. In the Hepatitis C case, relapse is produced after the end of the treatment. And in the Hepatitis B case, the biochemical and histological remission is only obtained in a third of patients. Is this obtained therapeutic benefit enough to justify the approval of Interferon in the Hepatitis treatment?. This question is intrinsically linked to the next one: Are there funds enough to afford this commodity?. Or even more precisely: Is Interferon, from the National Health Service's perspective, an efficient investment?.

Trying to solve this difficult problem, the Committee for Propietary Medicinal Products (CPMP) is looking for a generally accepted Summary of Products Characteristics (SPC) for the European Community before the end of 1990. This can partially answer the issue but several problems will still remain:

1) The indications demanded by the laboratories owners of the three Interferons are different. This means that, in the best case, we will have three different SPCs in Europe.
2) CPMP decisions are not binding at this moment.
3) In any case, each country needs to solve its financial problems through its NHS.

THE CASE OF ERYTHROPOIETIN

The introduction of recombinant human Erythropoietin has substantially improved the treatment of chronic anemias. It has been sucessfully used to correct the anemia of chronic renal failure in patients mantained by chronic hemodyalisis (11). Also this was the first indication that was approved throughout Europe using the Concertation Procedure that was established in the 87/22 EEC Directive. Afterwards, in July 1990 another indication was approved by the CPMP: "Treatment of severe anemia of renal origin accompanied by clinical symptoms in patients with renal insufficiency not yet undergoing dialysis". The approval of this indication, and the fact that there are new ones being studied such as the anemia of rheumatoid arthritis (12), anemic AIDS patients on zidovudine (13,) anemia associated with multiple myeloma (14) and even in patients with anemia of cancer (15), make a new therapeutic expectative, but once again limitation of resources to finance it must be recognized.

COST-EFFECTIVENESS STUDIES AND BIOTECHNOLOGICAL PRODUCTS

The expenses associated with Biotechnological Agents have a major impact on health-care economy. A global understanding of the overall impact of these drugs require a complete analysis not only about the cost of drugs but also about the associated changes relating medical-care cost and the resulting health benefits (16). The aforementioned problems with GH, Interferon and Erythropoietin, only could be undertaken from a global prespective if a cost-effectivenes evaluation is added to the classical regulatory evaluation.

It's not enough to decide what indications should be approved. But also economic evaluation must be carried out in order to assess if these therapies are good enough to be financed with public funds. The cost of Erythropoietin, for example, will not be offset simply by savings in transfusion costs. It is important, however, to consider all of the outcome changes that this drug will provide to the patient. These include changes in the cost of treating adverse effects and an improved quality of life (17).

Unfortunately, the methodology of these sort of studies still presents some difficulties and can't be systematically applied to every drug; although it is obvious that biotechnological products should be one of the first targets. It is probably premature to require it by the Regulatory Agencies, but seems judicious that health planners will keep it in their minds when economic resources are distributed.

CONCLUSION

This work started indicating that the quick development of products derivated from biotechnology has generated a conflict between four agents. By trying to find some advisable solutions, it is possible to suggest one for each.

- The developing Pharmaceutical Company. It will facilitate the resolution of the conflict if companies would not try to induce doctors to prescribe these products in indications that have not been approved yet. And also if they avoid conducting "seeding clinical trials" in this area, inducing the prescription of these products under the false appearance of a clinical research work.
- The potencial prescribing physicians. It would be useful if they refuse to participate in these trials, and they understand the fact that the publication of a clinical trial in a worthy journal does not always means that they must reproduce the proposed therapy in their daily practice until it is sanctioned by the Regulatory Authority.
- The Regulatory Authority. It would be useful a much faster reply to the discoveries obtained by clinical trials and if decisions taken would be detached from possible financing of these drugs by the NHS.
- The National Health Services. Their decisions about financing new therapies, especially expensive products like the ones derived from biotechnology, should be based as much as possible, on cost-effectiveness and quality of life studies.

REFERENCES

1. Raben M.S. J. Clin. Endocrinol 1958; 18: 901-03.

2. Milner R.D.G. Growth Hormone 1985 (editorial)
   Br. Med. J. 1985; 291: 1593-94

3. Luzuriaga T.C., Freijanes PJ. Utilización terapéutica de la hormona de crecimiento.Inf. Ter. Segur Soc. 1990; 14: 77-84

4. Rudman D, Feller AG, Nagraj H.S. et al. Effects of human growth hormone in men over 60 years old. N. Eng. J. Med. 1990; 323: 1-6

5. Internal Data (unplublished). Ministry of Health and Consumers Affairs. 1990.

6. Anonimous. El Consumo Farmacéutico en 1989. Inf. Ter. Seg. Soc. 1990; 14:73-74

314

7. Sánchez Franco F. Estado actual del tratamiento con hormona del crecimiento (unpublished data). 1990.

8. Davis G.L., Balart L.A., Schiff E.R. et al. Treatment of chonic hepatitis C with recombinant interferon alfa. N. Engl. J. Med 1989; 321: 1501-6

9. Di Bisceglie A.M., Martin P., Kassianides C. et al. Recombinant interferon alfa theraphy for chronic hepatitis C. A randomized double-blind, placebo-controlled trial. N. Eng. J. Med. 1989; 321; 9506-10

10. Perrillo R.P., Schiff E.R., Davis GL et al. A randomized, controlled trial of interferon alfa-2b alone and after prednisone withdrawal for the treatment of chronic hepatitis B. N. Engl. J. Med. 1990; 323:295-301.

11. Winearls CG, Oliver DO, Pippard MJ, Reid C, Downing MR, Cotes PM. Effect of human erythropoietin derived from recombinant DNA on the anemia of patients mantained by chronic haemodialysis. Lancet 1986; 2:1175-7.

12. Means RT Jr., Olsen JN, Krantz S.B. et al. Treatment of the anemia of rheumatoid arthritis with recombinant human erythropoietin: clinical and in vitro studies. Arthritis Rheum 1989; 32: 638-42.

13. Levine R.L., Englard A, Mc Kinley GF, Lubin W, Abels RI. The efficacy and lack of toxicity of escalating doses of recombinant erythropoietin (rHUEPO) in anemic AIDS patient on zidovudine (AZT). Blood 1989; 74: Suppl 1: 15a. Abstract.

14. Ludwing H. Fritz E., Kotzmann H, Höcker P, Gisslinger H, Barnas U. Erythropoietin treatment of anemia associated with multiple myeloma. N. Engl. J. Med 1990; 322:1693-9

15. Miller CB, Jones R.J., Piantadosi S., Abeloff M.D., Spivak JL. Decreased erythropoietin response in patients with the anemia of cancer. N. Engl. J. Med. 1990; 322:1689-92

16. Emery DD, Schneiderman L.J.. Cost-effectiveness analysis in health care. Hastings Center Rep. 1989; 19: 8-13.

17. Grimm AM, Flaharty KK, Hopkins LE, Mauskopf J, Besarab A, Vlasses PH. Economics of epoetin therapy. Clin. Pharm. 1989; 8:807-10.

Discussion -BIOTECHNOLOGICAL PRODUCTS, **CAN THIS COMMODITY BE**
AFFORDED?

P. Juul

I would like to ask you whether you would find it reasonable to combine the
registration authorities with the reimbursement authorities? I think that in most of our
countries these are two separate authorities which in many ways is an enormous
advantage for the authorities granting the marketing authorization. But with some very
expensive drugs the problem will evidently be that if the marketing authorization agency
accepts a new indication the reimbursement authority may be in a problem.

F. García Alonso

I believe this is the key question. The separation between the regulation and the
reimbursement. From a scientific point of view, the best way is to separate, absolutely.
But in the real practice this separation may not be evident. In the case of Spain, every
drug that we approve automatically is paid by the National Health Service and we cannot
resist the situation. We cannot pay for all the new drugs that have been approved.

P. du Souich

One very partial solution in Quebec is to approve the drug but with restrictions.
Then, in fact, only some specialists are allowed to prescribe the drug.

F. García Alonso

Of course this is a good solution but you need to deal with the legal peculiarities
in every country.

M.M. Reidenberg

In our hospital, and probably in many others, we have some reserved antibiotics
that require special permission to use, and then only for specific indications. The staff
has accepted this because of the issues of both cost and trying not to develop
resistance. We do have a model, probably all over the world, of restricting approved
drugs only to approved uses and indications, and I think this could be built on for other
expensive drugs if there was the will to do it.

F. García Alonso

This is a model that we can follow but, in real terms, at least in Spanish hospitals, we have difficulties to follow this system because of the resistance of the physicians. For example, we have restrictions on antibiotics but this approach often creates problems to the Commission on Infectious Diseases in the hospital. Biotechnology products are expensive drugs, and they are often overprescribed for indications that are not approved.

A. Ganser

I would like to say something from the viewpoint from a clinician. I see the limited resources that we can spend for drugs. But what I cannot accept actually is that we should not look into the journals and see whether new applications of drugs have become available. And if there is a publication, even in one of the major journals, I think we have a responsibility as clinicians to see whether this actually holds true. What we should do, I think, is to limit the use of these biotechnology products to certain doctors, as it is done in Quebec.

F. García Alonso

But the question remains: who is the authority in charge to say who are the physicians that can prescribe and who are not? This is really very difficult because every physician mainly working in the hospital, believes that he is a very important researcher.

J. Bigorra

I believe that in general terms therapeutic advances should be made available to society. The question is who will pay for it? I think that, in principle, in a private relationship between a physician and a patient, the patient could pay and there would be no problem. However I also understand that if a new principle needs an experience of more than 20.000 patients to establish what would only be a very minor advancement, the National Health Service can decide that this is not of interest for full reimbursement.

B.R. Meyer

In regard to Dr. Reidenberg's suggestion. We have such a policy where we have very strict restrictions on erythropoietin, intravenous gamma globulin and interferon. And while those are very effective in ensuring that the drugs are only used in appropriate situations it does not ensure that they are cheap. For instance, we only approve erythropoietin for renal insufficiency or renal failure with anemia and intravenous gamma globulin is approved for ITP, congenital agammaglobulinemia or AIDS patients with severe hypogammaglobulinemia. Despite that and being sure that 98% of usage is appropriate, we still have very extensive costs. I do have a hesitation about restrictions because I do think that therapeutic experimentation of physicians who have a unique case and a unique problem and who attempt to develop a unique solution has been a source of consistent innovation in therapy. So that I also think there is a problem, aside from the obvious ones, which is that restriction tends to rigidify our attempts at therapy and the person who is trying to come up with a novel solution to a difficult novel problem is appropriately angry and frustrated and I think in the long run this may hurt us.

W.M. Wardell

What we are dealing with are systems for controlling drug utilization. Such systems exist and have existed in various forms in many countries for years and are inexorably, for better of worse, encroaching on the use of pharmaceuticals in most countries. The systems can be tuned through computer diagnostic criteria and computer reimbursement lists to just about any degree of constraint that the payor desires. The central question still is once the drug expenditures of a country have been drastically cut by such mechanism, who is going to fund the R&D for new drugs? It is a real dilemma. I come back to the thought that the only way to ameliorate the situation, looking ahead, is to reduce the costs of development and lower the barriers to approval. Once you take the incentives away, new drugs won't be developed.

F. García Alonso

But of course, there is no evidence that reducing the development costs would automatically reduce prices.

W. Wardell

But increasingly, the prices are in the hands of the managers of utilization control systems.

CONCLUDING REMARKS

MARCUS M. REIDENBERG, M.D.

Departments of Pharmacology and Medicine, Cornell University
Medical College, 1300 York Avenue, New York, NY 10021 USA

This Fourth Esteve Foundation Symposium was planned to bring
together researchers in the new biology, clinical pharmacologists, physicians and administrators to share experiences and
think together about how to convert the potential of modern
biological science into the reality of improving the health for
all people. During the course of the symposium, in both the
formal presentations, the formal discussions reproduced in this
volume, and the informal discussions held at mealtimes and late
into the night, several themes and issues emerged that are worthy
of highlighting.

First with respect to the drugs themselves, several ideas were
repeatedly articulated. Primarily was the fact that the use of
peptides and proteins as drugs has a long history and is not a
recent innovation. Some examples include vaccines, whole blood
transfusion, plasma and selected plasma proteins, ACTH, growth
hormone, insulin, etc. Thus, while the process of obtaining the
peptides and proteins by the techniques of the new biology is
new, the development and use of peptides and protein molecules as
drugs is associated with a long clinical experience and is not
new. Two thoughts emerge from this realization. Firstly, many of
the newly available peptides and proteins really function like

hormones so that proper dose-response and time-action studies have to be done early in the drug's development in order to determine the proper dose and route of drug administration for the therapeutic clinical trials. Secondly, the development of antibodies to the recombinant proteins with adverse consequences for the patient was a real concern early in the development of recombinant proteins. Or the basis of experience with a number of new drugs as well as old drugs like insulin, this problem of antibody formation leading to adverse clinical consequences seems to be much less of a problem than originally feared.

The second major theme was concern with establishing appropriate standards for safety assessment of biotechnology products as drugs. Fermentation broths that are the starting material for recombinant proteins and peptides contain an enormous number of other substances that must be removed in the process of converting a broth into a medicine. The issue of how pure a product is necessary requires a rational scientific answer. As increasingly sensitive analytical methods are developed for testing for compounds known to be present in fermentation broths, small residual amounts can be detected as impurities in the final product. All of these tests have not been carried out on traditional biological products like vaccines, insulin, or purified plasma proteins. Thus, the long experience with the traditional biologicals is not of use for assessing the importance, if any, to the levels of many impurities in biotechnology products. It would appear logical to determine the maximum level of an impurity that would be acceptable based on biological facts and principles and not on the analytical sensitivity of the current method of assay. Then that level should be set as the standard. In addition, the acceptable levels of some of these substances, such as certain cytokines, could vary if the product under consideration was to be used once in the course of an illness rather than repeatedly or chronically.

What constitutes an appropriate preclinical safety assessment for biotechnology products was another area of discussion. Much of what is currently done appears to be based on customs that have been developed for the toxicologic assessment of small organic molecules. Thoughtful evaluation of the potential toxicity of biotechnology products would lead to the questioning of some of the current practices. For example, some of the recombinant proteins are highly species specific. The human proteins do not produce a pharmacologic effect in the common laboratory animals. They do induce an antibody response. Thus, the usual preclinical test of daily drug administration for months to rodents or dogs cannot indicate what the pharmacologic effects of chronic drug administration to man will be and may lead to an immune-mediated disease in the animal that has no predictive value for human risk. For drugs like this, traditional preclinical multidose toxicity testing is not predictive and can be misleading. Certainly, the utility of these and the rest of the preclinical toxicology assessment should be continually re-evaluated in the light of developing knowledge. Continuing to do tests without predictive value and with the potential for misleading results makes no sense at all. They only add to the cost of the final product.

In this regard, discussions highlighted the fact that there are differences between the preclinical safely assessment requirements of the regulatory authorities of the United States, Japan, and some of the countries of Western Europe. No scientific basis could be expressed for why preclinical assessment requirements should differ among countries. The development of biotechnology products could be facilitated and the cost of development reduced by the various regulatory authorities agreeing on a single set of rational scientifically-based preclinical safety assessment requirements that would be acceptable in all countries.

Another aspect of regulatory requirements that drew comments was the idea expressed by some authorities that following process changes in the preparation of a biotechnology product, new toxicology or clinical studies should be done. An opinion was vigorously expressed that if the product prepared by the new process meets all of the specifications established for the product, then requiring testing of this product that was not a part of the batch testing requirements for the product prepared by the old way simply added cost without adding benefit.

Another issue related to regulation but much broader than regulation alone is that of using surrogate endpoints. By surrogate endpoints, one means endpoints that are more easily measured than the really desired endpoints and are thought to be predictive of the desired endpoint. Surrogate endpoints are accepted in some therapeutic areas in which there is great confidence that the surrogate endpoint predicts the desired one. For example, antihypertensive drugs are evaluated on their ability to lower high blood pressure rather than to prevent strokes even though stroke prevention, not lower blood pressure, is the really desired endpoint. More problematic may be the answer to the question of what really constitutes an appropriate endpoint. An example came out in the discussions of thrombolysis. Is re-canalization an appropriate endpoint for a trial of a thrombolytic drug or is survival the real endpoint with re-canalization being only a surrogate endpoint? A different example came up in the discussions of the hematopoietic growth factors. Is a decrease in infections in patients receiving antineoplastic therapy an appropriate endpoint itself or is it a surrogate endpoint for the real endpoint of prolonged survival? If decreased infections is only a surrogate endpoint, what research must be undertaken to evaluate whether it is predictive of prolonged survival?

Finally, two themes emerged from the discussions of some of the economic aspects of the development of biotechnology produces. One was the high cost of developing these products which is one of the contributors to the price of the products. Questions about how much useful information is gained from some of the preclinical testing were also raised. An issue discussed was how much information about efficacy and safety is really gained during the present phase III trials, especially those parts of the trials that simply increase the numbers of patients on

protocols that have already been carried out in modest numbers of
patients. One can legitimately ask, What is the minimum information necessary to assure efficacy and safety and safeguard the
health of the public? Then one can ask, Is it reasonable to
require more than this for registration?

Another aspect of this issue is the whole concept that a drug
is purely experimental on one day, is approved for marketing by a
regulatory agency on the next day, and is safe and effective on
the following day. Clearly, this is not the way information about
drugs is obtained. There is more unknown about a drug earlier in
the development process than later. Everything about the drug is
not known at the time of registration, or a year later, or even
ever for that matter. Information about a drug is accrued
gradually and continually. But our regulatory processes are not
designed to take this into account in the registration procedure.
In my opinion, phase III testing prior to registration is too
prolonged and extensive. This long and extensive testing delays
the general availability of good new drugs while contributing
little to the assessment of safety and efficacy beyond what was
learned in phase II and early phase III. The reason is that
going from several hundred patients to several thousand contributes little additional power to detecting uncommon adverse events
compared to the increase in drug exposures to hundreds of thou-
sands or millions of patients after marketing has begun. Since
no phase III would have been started unless efficacy had been
found in phase II, little, if any, important efficacy data is
generated by the large scale phase III trials. Some specific

targeted studies to answer specific scientific questions remain after phase II and are done during phase III. There is a real question, however, about whether the incremental information gained by just adding numbers of patients beyond some minimum number (depending on the drug and disease) contributes to the health of the public or impairs it by delaying general availability of a good drug and by increasing the development cost and thereby increasing the price of the drug to the patient.

The time unusual events are likely to occur is after marketing when the number of exposed patients can be increased by several orders of magnitude. Yet it is very hard for a regulatory agency to take a regulatory action after registration has occurred. It seems to me that society would benefit by substantially reducing the present amount of phase III research prior to registration, implementing a valid postmarketing surveillance program, and making it easier for regulatory authorities to regulate registered drugs including revoking the registration if problems occur postmarketing that would have prevented registration if they had been detected prior to registration. One could get evidence for the validity of such a change in policy by learning how many new chemical entities failed to gain regulatory approval because of safety concerns that were initially discovered by the large phase III trials, how many of the drugs recently withdrawn from the market because of safety concerns should have had these problems adequately identified during the present phase III practices and why they did not, and the extent to which the marketing of important drugs has been delayed by present phase III requirements. A thoughtful review of potentially available information should be able to determine the wisdom and the risks of a change in regulatory policy to that suggested.

All of the participants in symposium hope that their deliberations, as reported in this book, will contribute to advancing the knowledge about biotechnology products and facilitating the translation of biotechnology into improved health care.

Index of Authors

Keyword Index